AF820839

SOTHIS PRESS

Sothis Press aims to completely reimagine the public domain book with an emphasis on universal access to captivatingly designed print editions. We believe that technology will enable every single book ever published to be perpetually accessible in beautiful print form. Sothis is beauty in ubiquity.

www.sothis.press
1 Waterhouse Square, London, ECIN 2ST

Cell Intelligence

The Cause of Growth, Heredity and Instinctive
Actions Illustrating That the Cell Is a Conscious,
Intelligent Being, And, by Reason Thereof,
Plans and Builds All Plants and Animals
in the Same Manner That Man Constructs
Houses, Railroads and Other Structur

Nels Quevli

SOTHIS PRESS

ISBN 978-1-0365-2687-0

This edition is published by
Sothis Press © 2023. The company
is registered in England and Wales.
The Company Number is 14900389.

CELL INTELLIGENCE

the Cause of Growth, Heredity and Instinctive Actions, Illustrating that the Cell is a Conscious, Intelligent Being, and, by Reason Thereof, Plans and Builds all Plants and Animals in the Same Manner that Man Constructs Houses, Railroads and Other Structures.

By

NELS QUEVLI

REG. PHAR., LL. B.

THE COLWELL PRESS
MINNEAPOLIS, MINN.
1917

CONTENTS.

This Book
is lovingly dedicated to
my wife
Anna MaGuire Quevli

PREFACE

The purpose of this book is to introduce you to your maker, the Cell, to get you better acquainted with him and to let you know that he is an intelligent being and very likely more so than yourself.

The proposition that the cell is your maker or builder, that he is the cause of and builder of all plants and animals and that he is a conscious and intelligent being is a broad and sweeping statement. I do not think it has ever been made before in the history of the world. This proposition will no doubt be hotly contested by those institutions who may think that they will be financially affected by these facts becoming general knowledge.

Someone said that "The greatest study of mankind is man." I would say, the greatest study of mankind is his maker, the Cell. This book will explain to the reader the cause of evolution, or growth, heredity and instinctive action in plants and animals. It will show that all plants and animals are built and produced by the microscopic beings we call cells. It will show that in their place in life they exercise the same intelligence in reference to their work as we do in ours and by reason of their intelligence, they are able to build a plant, a tree, an insect, animal or man, the same as we are able to build a house, automobile, ship or railroad. I propose to show by this book that although microscopic in size, the cells are no less intelligent than we are. It will show that the cell knows how to multiply and organize his offspring into the vast co-operative colony we call plant or animal. Why should the builder and organizer of a large colony or army be less intelligent than the army? In other words, why should you be more intelligent than your maker? This book will explain the problems:

1st—Why will an acorn grow into an oak?

2nd—Why will it grow into an oak and not into a maple?

3rd—How can a bird raised in captivity build a nest like its parent when it has never seen a nest before nor ever seen its parents?

This book necessarily covers a wide field of knowledge from the fact that it attempts to show a conscious intellect in the cell similar to our own. In this inquiry it was necessary to investigate biology in all its branches, viz: embryology, heredity, psychology, physiology, anatomy, zoology, botany, and also chemistry and physics to a certain extent. The cell being the cause of all living structures, it was necessary to investigate his activities in all places. The plants and animals are cell colonies and are produced and maintained by them for their own selfish purpose. While I have tried to make the work as scientifically accurate as possible, I have also tried to avoid all technical words and phrases, so that the average man could understand it.

While the opinions set forth in this book are based on my own investigations, I have also given extracts very liberally from others, so that the reader can better judge for himself on disputed points. I have introduced a large number of illustrations in order to make it clear, to simplify, and to save time for the reader. I have not overlooked the great difficulty with which I shall meet in order to be able to get this book before the attention of the average man, as I shall probably be attacked by all institutions who may think they will be injured in their business. All colleges controlled, (and it seems that a large number of them are so more or less) will join with others in criticizing this book. No other branch of science has been so willfully obscured and mystified as the subject of life and especially the life of the cell. Even today if you try to discuss or explain the cause of a seed growing into a tree to the average so called educated man, he will only answer with an incredulous smile. He seems to know little or nothing of the life of the cell, which so vitally concerns the health and happiness of every person. I can only account for this condition of things by the willful suppression of the truth by the controlled institutions of learning. Such men as Chas. Darwin and Ernest Haeckel have done as much towards the advancement of human knowledge and progress as any two persons that have ever lived.

Health is impossible without a clear understanding of life and without health, a happy life is impossible. For that reason I consider those two great students of life among the greatest benefactors of mankind.

Dated, March 1915. NELS QUEVLI.

LIST OF ILLUSTRATIONS.

NELS QUEVLI

Cell Intelligence the Cause of Evolution

CHAPTER I.

INTRODUCTORY.

The purpose of this book is to clear up some of the mysteries of life, such as the cause of growth, evolution, heredity, instinctive, impulsive and reflex actions in man, animals, and plants. No one seems to understand or comprehend their real cause. Why and how does a certain seed grow into a weed or grass and not into a tree? Why do the young of animals, including man, look and act like their parents? Why does a kernel of corn develop into a corn stalk when placed in the ground and not into a sunflower? How can a bird raised in captivity, who has never seen its parents nor ever seen a nest of any kind, build a nest just like that of its parents? It knows how just as if it had been fully trained to do so, although it has not had the slightest experience. People do not seem to know, and call it instinct. I have read everything that I can find on the subject and writers all seem to consider it a mystery. Now as it is no longer any mystery to me, I feel as though it might be my duty to write this book and give my opinion of it for whatever it is worth.

The textbooks of science have become very complex and technical and very numerous, so much so that a person in the ordinary busy life has not the time nor inclination to attempt to read up and study them so as to understand the great questions of life. The broad realm of science has been greatly extended in the 19th century. A great number of branches have been established by reason of research in different directions. This enormous expansion of the field of knowledge has led to a narrow specialism in many small sections. The extensive division of labor has prevented a comprehensive consideration of life as a whole. Each man has kept his mind on his little separate proposition and invented a lot of terms, words and phrases, which he has used in his little one sided work. This has caused a great deal of confusion and misunderstanding. One of the greatest scientists in Germany, Ernest Haeckel, states: "The vast structure of science tends more and more to become a tower of Babel in the labyrinthic passages of which few are at their ease and few any longer understand the language of other workers." This statement certainly is true and that is one of the main reasons why this book is written.

There is no reason why a majority of the people should spend their lifetime upon this planet and know nothing or little about this interesting question of life and what it is. In this book I will try to explain my ideas in the very simplest language possible. The language used in almost all the text books is too technical and beyond the reach of ordinary readers. For instance—to give the reader an idea of what the subject of life includes, and what I have been reading, I shall outline a few subjects.

First—This whole subject of life is called biology. This includes protistology or the science of single cells;

botany, the science of plants; zoology, the science of animals; anthropology, the science of man; morphology, the science of forms; physiology, the science of functions; anatomy, the science of structure; biogeny, the science of development; aecology, the physiology of work; histology, the science of tissues; organology, the science of organs; blastology, the science of persons; and cytology, karmology, phylogeny, palaentology, geneology, trophonomy, chorology, ontogeny, sensonomy, psychology, chemistry and physics. However, after reading these subjects, you have a lot of theories about metaphysics, mechanism, hylozoism, dynamism, idealism, materialism, hedonism, monism, dualism, vitalism, gases and fluids, statics, dynamics, acoustics, gravitation, electricity, and a great number of others too numerous to mention.

Now it is clear that the busy man of today cannot afford to read up on all these subjects and theories, even if he understood the language used, as it would take up half of his life time to do so. After reading up on these subjects, I do not agree with any of them as to the cause of growth, evolution and development of life, heredity, instinctive and reflex action, and that is why I have written this book.

After I have stated what my opinion and ideas are on these subjects, the reader must not stop and say to himself,—"Why, that proposition is ridiculous, I do not believe it," but he must go on and read the whole book, as this is a big subject and no one can understand it so as to form an opinion without reading every chapter of this book. For thousands of years we thought the sun coursed around the earth; it was a bold theory to bring the sun to a standstill and set the earth in motion. We remember how poor Galileo had to suffer inprisonment for supporting this theory, which had been proven seventy years previously by Copernicus. The astronomers were able to bring forward so many facts to support this theory that its truth was finally forced upon every thinking person.

The main fact I propose to prove by this book is that the cell is an intelligent being like ourselves. I have stated my proposition and it is now up to me to prove it. The proposition that the being we call cell is an intelligent being, includes also that he is so in every sense of the term, that he is conscious, has a memory, will and judgment, just as we have. He reasons and profits by experience, just as we do. By virtue of his power of memory and experience had in our bodies and thousands before ours, he knows how to build others like them.

Before I go any further I must quote a few pages from a physiology now used in the high schools and which most of you have studied, so as to fully remind you of what is known of this important individual we call cell; that is, that he is an independent living being; that he also lives a separate life, just as we do; that he has the power of growth, self-repair and increase in numbers; and that he requires food, air, water, and shelter, just as we do. Here is what our physiology has to say about the life of the cell: "There are certain great differences between this unit of living matter and an unliving thing. Three distinguishing qualities belong to the living cell. (1) growth; (2) self repair; (3) increase in numbers through self division. These powers are possessed by no other material in the world save protoplasm.

"The growth of a cell is in all cases brought about by material taken in from the outside. In the human body this material is food, which after digestion passes into the blood, and is then taken in by the cells. This process will be described more fully later. In some of the very lowest organisms where the whole animal is a single cell, solid particles may be taken into the cell through definite openings or "mouths." In others, the cell may change its shape so as to wrap itself about the particle to be taken in. But even in these instances, the particles must be dissolved or digested before they can be built up into the protoplasm of the cell.

"At the beginning of its life, the animal consists of a single cell only, an egg, but as it grows the number of cells increase. When a child grows to manhood his increase in size is not due to growth of the individual cells making up his body but to the increase in their number. The cells of the adult are not larger than those of the child but more numerous. This multiplication is the result of repeated division of the original cell and in this process, every part of the cell divides. . . . . Each of the resulting cells immediately commences to grow and continues until it becomes as large as the one from which it started, then it divides and the story is repeated.

"We have thus seen that the body is made up of organs, that the organs are made of tissues and that the tissues are made of cells. Is it possible to carry this division further or is the cell the final unit? To this question we must reply that the cell is the final unit. It is true that the cell has parts, cell wall, nucleus, cell substances, etc., but no one of them can live by itself, while a complete cell may be an independent body and live an independent life. Although our own bodies are composed of many millions of these cells, there are some organisms made up of one cell only—these are microscopic and are called unicellular animals and plants. Although very tiny, each lives an independent life. . . . . . . They vary in shape and differ in structure. Some of them have "mouths"; others simply take their food in at any part of the body by allowing their protoplasm to flow around it. Some of them have organs for locomotion, others do not; some have shells, while others have no covering at all. But each is a single cell and each carries out its own life process, such as respiration, secretion, multiplication, etc. The cell cannot be subdivided into smaller units which would be able to sustain independent life. . . . . .

"Since such cells are the simplest parts into which living matter can be divided, we may call them "Units of Life" and may regard our bodies as a combination of a large number of such units, considering the life of the whole body as the sum of the lives of its different cells. We should constantly remember that it is really the cells which are the active living parts. The combined lives of all these millions of cells make the life of the whole, much as the combined lives of the persons within a city make up its life.

"As we have seen, some animals are composed of a single cell. But this cell is able to carry on all the functions of life: it feeds, digests, respires, moves, multiplies and performs all the necessary duties of complete, individual life. In our own bodies there are many cells, but each is not capable of carrying on all the functions of life and if separated from the others, would die. Each is able to do primarily only one thing, hence, each is dependent upon the others.

"It may be asked, why we should have so many kinds of cells in our bodies and why with us too, one kind of cell could not serve all purposes. The answer is easy to give. A hermit can himself do everything needful to support his life: he can prepare his own food, make his own clothes and build his own shelter, but he can do this only because he lives very simply. When a family lives alone on the frontier, the members divide the work among themselves; the husband doing the work out doors, the wife that indoors, and the children contributing their different shares. When several million come together, it will be found that some members of the community are more skillful in building houses, others in making shoes, others in dressmaking, still others in cooking, etc., so the people agree to divide their tasks and share the results of their work. In this way they may have better houses, better shoes, better clothing and better food than before, because each man does what he can do best. As the community grows, this division of labor becomes extended until in a large city each person does only a very small part of the work necessary to supply him with the things he needs. But he can do his own work well because he has only one thing to do. The life of a city is of much higher grade than that of a pioneer family. Its population has many more luxuries and accomplishes much more, all because of this division of labor. So it is among organisms. Where one cell does everything, the life is simple and on a very low scale. Each cell can feed itself and perform all the necessary functions, but the whole life is only one of growth and reproduction. As the cells become more abundant, they also become alike. Each takes upon itself certain duties; each contributes to the good of the other cells and each receives aid from the others. . . . . The life of any animal is the sum of the lives of its cells, and with many kinds of cells all working together for a common good, a higher grade of activity is produced than with each working for itself alone. Division of labor goes hand in hand with a rise in the scale of accomplishment and results in a superior type of life."

Ernest Haeckel states, "Cells are grouped together under the idea of sculptors or builders because they alone in reality build the organisms."

I go into details as to what the cell is, presuming that the reader has not any previous knowledge of what the cell is understood to be. I think it is necessary to do so in order to make a scientific subject like this interesting to the reader who may not know what the cell is. This you will see makes it clear that this animal or being we call cell is the agent who builds or makes the things we see in life, which we call plants and animals, including man. These wonderful beings are too small to be seen with our naked eye. They live separate and independent lives in both fresh and salt water. Such words as protists, copepods, crustaceans, protoplasm, bacteria, germs, leucocytes, histons, somatic cells, germ cells and amoeba all mean the same, that is, they are cells. It is admitted by all that all living things, plants or animals, that we see are produced by these beings. The next question then is this: Do they do so by reason of their intelligence or do they do so by reason of some chemical force or otherwise? My proposition to be proven by this book is that they do so by reason of their intelligence in the same manner as we build a house or railroad. We are not prepared to say at this time just what intelligence is except that we know it exists. We know we are intelligent and are guided by intelligence in our acts. These beings when they live separate lives in the water act precisely as other animals. They show, I believe, by their acts the same intelligence in their place in life as we do in ours.

Looking at them through the microscope, Mr. Binet states: "If a drop of water containing Infusoria be placed under the microscope, organisms are seen swimming rapidly about and traversing the liquid medium in which they are, in every direction. Their movements are not simple; the Infusoria guides itself while swimming about; it avoids obstacles; often it undertakes to force them aside; its movements seem to be designed to effect an end, which in most instances is the search for food; it approaches certain particles suspended in the liquid, it feels them with its cilia, it goes away and returns, all the

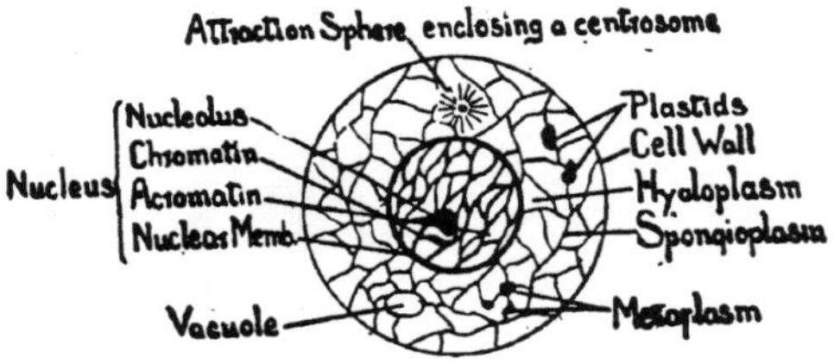

Fig. 1.—The Cell

while describing a zig zag course similar to the paths of captive fish in the aquariums; this latter comparison naturally occurs to the mind. In short, the act of locomotion as seen in detached Infusoria exhibits all the marks of voluntary movement." In this statement, Mr. Binet called them Infusoria. This animal has organs and acts precisely in the same manner as any other animal. It feels objects to decide whether they are good for food. It has a will, judgment and discretion, just as we have. It can decide for itself whether any particular substance is fit for food. It can exercise discretion in the matter in the same way that we do in hunting for something to eat.

The scientists of today do not consider the cell an intelligent being. They claim that he does not show anything but irritability. This word has caused much confusion and is practically meaningless as far as I can see. No one seems to know what is understood by this word. Mr. Binet seems to be also of the same opinion and states: "The term irritability, which though long in use has not in our opinion been happily chosen, since it is in the highest degree ambiguous and not suggestive of an exact signification."

The little fresh water cell we call amoeba, who leads a separate independent life, is so very similar to those that build animals and man that the microscope can discover no difference. Mr. Binet described his actions as follows: "The following is what occurs when the amoeba in its rampant course happens to meet a foreign body. In the first place, if the foreign particle is not a nutritive substance, if it be gravel for instance, the amoeba does not ingest it; it thrusts it back with its pseudopodia. This little performance is very significant; for it proves, as we have already said, that this microscopic cellule in some manner or other knows how to choose and distinguish alimentary substances from inert particles of sand. If the foreign substance can serve as nutriment, the amoeba engulfs it by a very simple process."

You will see from this that his actions show discretion. In hunting for food he must exercise his judgment at all times as to where he is going and as to whether this or that is suitable for food. Such action of choosing must be based on former experience, which involves memory. Again here is what Mr. Binet has to say about the actions of cells, showing how they act when living as separate beings and not in those vast co-operative colonies we call plants and animals: "The Infusoria when in a medium abounding in food are almost entirely sedentary in their habits, only making slight changes of position. But if they are placed in a medium affording but little nutritive material, they become as migratory as the hunters and are seen to race about in all directions, searching for more abundant nutriment. It is hard to find a more perfect illustration of the influence exerted by the condition of a medium upon the habits and customs of animals. The Patula is a type distinctively carnivorous and possessed of an extremely voracious appetite, a fact which explains its power of multiplication, one of the greatest I have studied. . . . . In constant pursuit of its prey, it seizes its victtim by the two stout vibratile lips with which its mouth is armed and swallows them alive and whole. The victims may be seen struggling and tossing about for a time in the interior of its body and afterwards to expire slowly under the action of the digestive juices of the vacuole in which they have been inclosed."

Notice in this how precisely similar the actions of these cells are to those of animals and human beings. They lie around and do nothing towards finding food as long as they have enough to eat, but lack of food and hunger stirs them to activity. The actions of these cells constantly in pursuit of their prey and the manner of eating and digesting their victims are certainly significant. They go to show that the animal or cell has the mind to know what he wants and goes after it. It is impossible to conceive of any living being moving towards an object for a purpose which has not a will and memory. It never can be shown to be a chemical or mechanical act. There must be in the mind of the being the feeling or idea of a need of food to spur him to action. Chemical force and other natural forces always act the same and follow fixed laws regardless of any condition or need. All scientists agree that matter can only act and change its place and form according to fixed laws. Living beings act according to their wants and needs. They are masters and are able to direct the blind forces and crude matter to their own purpose and use. It may be said that even if it be proven that the cause of evolution, heredity and instinctive action is the result of the acts of the cell and that his acts are guided by intelligence, can you explain what intelligence is? That question may not be satisfactorily explained at this time; however, we all can form a fairly good idea of what it is from our own experience. After we have fully proven that it is by reason of his intelligence that the cell is able from memory and experience to build the different structures we call animals and plants, then we can begin to push further investigations to discover, if possible, the deeper secrets of the inner life of the cell.

It is a rule of science and philosophy that we should investigate a proposition as far as we can and when we can go no further to stop and admit it. No one will deny that there is such a force as electricity. We can see and feel its effects, but as yet no one is able to say just what it is although some very good theories have recently been advanced. It is the same with this property we call intelligence, that guides us in life. We know about what it is by the way it guides our actions. It is enough to know for the purpose of this discussion, that it does exist.

There are two main theories by which the growth and development of plants and animals in life are explained: First, chemical and mechanical forces; second, Intelligence or a Divine Being. However, so far no one has yet ventured the proposition or statement that the intelligence that has caused the production of all these structures we see, such as plants and animals, was the property of the cell. Some call it God's will in the universe or Divine Wisdom. Some call it intelligent force or vital force. They see a purpose and design in nature which seems to prove to them conclusively that it is guided by intelligence, and sometimes it is used to prove God's existence in the universe or a Divine Will or the will of some other intelligent force in the universe. The other side attempts to explain the growth and development in life from chemical, mechanical and natural forces. Now whatever you wish to call it, not one can deny that whoever the being is who is able to effect the construction of such stationary structures as trees and plants and movable objects as animals, must be possessed of a very high degree of intelligence. They are structures that present themselves to us with all the characteristics of a work of art. It stands practically admitted, that the cell is able to produce all these structures, partly of himself and partly from the crude elements of earth, air and water. We will see later when we get to it that he gathers the materials with which he is able to effect these constructions in the same manner as we do. That in order to be able to manufacture the different materials that are needed to do this, he takes advantage of heat, light, electricity and chemical actions in exactly the same manner that we do except, of course, on a much smaller scale. You know that so far we have not been able to find anything in crude matter and the blind forces of nature indicating any intelligence. Nothing is found indicating any power to produce a work of art. It has now been conclusively shown that the cell is not a simple organism but a very highly organized animal or being, which is made up of still smaller beings. For convenience in the discussion of this question, we shall call this being the primordial cell. It begins to look as if these primordial cells or beings of which the cell is composed are of that microscopic size that they may be able to handle the molecules and atoms of matter and direct their actions and motions in the same manner as we are able to handle brick, stone, and larger particles of building material. We will see later that these primordial cells possess intelligence; at least, their actions in the cell so indicate. You see, we are getting down to the question: Does intelligence exist in the atoms and molecules of matter or is it a separate thing from matter existing in the universe? Is it invisible force separate from matter similar to the force of electricity or magnetism? These are questions to be settled by the scientists in the future. The only purpose of this book is to show that the cause of plants and animals, as we see them, is intelligence in the cell. He is the being who does the work of building and constructing all these living things, and he is able to do so by reason of the fact that he is an intelligent being like ourselves. We are quite conscious that there is a very great deal concerning which we know little or nothing. For instance—we have practically no idea of the nature of either gravitation or electricity. We realize that we have yet much to learn. We do know that such things as gravitation, intelligence and electricity exist.

Memory is the power to receive impressions of things,—keep them for a long time and reproduce them when needed. We know this faculty of memory to be the base on which our intelligence rests. Without this power to remember and compare experiences, we could form no ideas or judgments. We would be unable to do anything based on past experience; in fact, there cannot be any intellect or mind without memory. By reason of this power the cells can remember just how the structures were made from which they came, and how they acted while a part of such structures. The cells build the new structures, such as plants and animals, as like as they can remember, from their experiences in the structures from which they came, and also from their experience in building similar structures in ages past.

The instinctive emotional and reflex actions are generally those produced and directed by cells in other places of the body and not in the brain. All these actions will be considered and explained in the chapter on heredity.

You have probably been wondering what I would have to say about the theory of evolution being the cause of life. That theory does not exclude but goes to support my theory. Evolution is an undisputed fact. In a struggle for existence the fittest will survive, but the fact is only an incident in life. Survival of the fittest is a fact in nature, in business, politics, religion and among nations. In the struggle between two battleships, the best fitted for the struggle will generally win out. My contention is that the real cause of the existence of those individual struggling ships was the intelligence possessed by the builders. The ships did not come to exist by chance arrangement of matter. We shall find upon investigation that natural selection and survival of the fittest are the things that determine who shall live of two structures or individuals contending for space and existence. In the same manner, survival of the fittest is the thing which will determine who shall be the ruler of the sea, the English or the Germans, or whose factory shall survive in a town where only one can exist. The theory of evolution propounded by Darwin and his followers does not account for the existence of the German and English navies nor for the existence of the factories, who struggle for existence, one against the other.

My contention is this, that these factories and navies did not come to exist by chance. Some intelligent being has caused their construction or existence. Who are the builders? Are they intelligent beings? We must ask the same questions in regard to the structures we see in life. We see a tree. It is a stationary habitation of a billion cells. We ask, who are the builders of that tree? It is admitted now that the cells are the builders of the trees they occupy. You ask, how can they build them? They do so by gathering the material and arranging it into the form of a tree. How could they do it? Because they are intelligent beings. The reader may not be familiar with Darwin's theory of evolution, so I shall briefly state what it is:

First. All animals vary. Hardly ever are two born just alike. There is nearly always some point of difference.

Second. All animals and plants multiply so rapidly that all of them cannot possibly live.

Third. All animals and plants are therefore in a constant struggle with each other for food and existence.

Fourth. Those who have by chance been born with some feature that favors them in their struggle will live and perpetuate their kind.

Fifth. This favorable variation will be passed on to the offspring through inheritance and in that way preserved to that species.

Sixth. By this process of continual variation, struggle for existence and inheritance, the different species of animals have arisen. Darwin has also added sexual selection. He shows that among many animals, there is a contest in the breeding season among the males for the possession of the females. The contest is sometimes by actual combat and sometimes by an attempt to attract the favor of the female by the display of brilliant plumage or by singing. Now evidently those males who have especially developed weapons for combat or especially beautiful plumage will be the victors, and will leave the most vigorous offspring, and the offspring will have the tendency to inherit the same weapons and plumage. This process developes antlers on the male deer and elk, and the beauty of the male birds. By means of this principle Darwin and other evolutionists have attempted to account for the evolution of life and the origin of new species from the old upon purely physical laws.

We shall see that this theory is true only so far as it goes, and shall also soon see that upon analysis it does not go any further than I stated. That it is merely an incident and a fact that the best man or institution in the struggle for existence will win and live to perpetuate his kind or his institution or business. It does not explain who are the builders of these struggling institutions, structures or beings, nor how the builders are able to put them together. They do not attempt to explain or give any cause why one should inherit the features, form and character of his parents, nor why individuals should vary. In this book I shall attempt to definitely settle these questions, so that we know where we are at, and can start investigations of other and further questions. As Darwin's theory of evolution is based on inheritance, it must fail as a cause, as a great number of ants and bees have neutrals, that is, they have males, females and workers. These workers can in no way pass on anything to their offspring, as they never have any. These matters will be fully discussed when we get to them. Spencer, like Darwin and others, attempt to show that life is based on chemical and mechanical action. They fail, however, to explain the mystery of growth, heredity and instinctive action. To show you how hard they have tried to understand it and how they have twisted the language to say something that would sound like a cause I shall here quote some passages from Spencer, the great exponent of the Darwinian theory. He puts the question: "How can this cell construct a body?" Does every germ contain a model of the complete animal or is each germ a miniature form of same? No, all we can say is that the living particles composing one of these fragments have the innate tendency to arrange themselves into the shape of the organism to which they belong. He seems to see no other reason for the actions of the cell in building a plant except that they have "innate tendency" to do so. He would explain it just as well by frankly stating that he did not know.

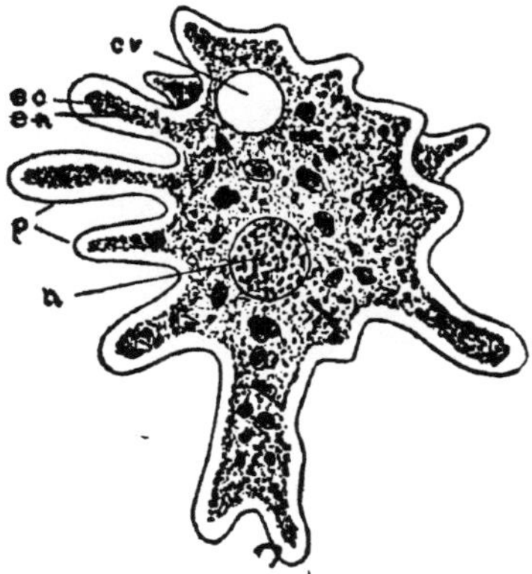

FIG. 2.—Amœba highly magnified—Central portion, cell body. n. nucleus; c. v. bacuole; p. Pseudopodia, hands or arms extended.

The question that will be discussed by me is not what intelligence and vital force are, but whether or not this intelligence that directs the cell is similar to and the same as that which directs intelligent man. Many scientists today are studying questions and theories which exist only in the imagination of the person who is studying them. We shall take, for instance, the subject of heliotropism of living organisms. This word is used to denote the property in some plants or animals which compels them to go towards the light. I have examined this subject and I am certain that no such property or thing exists; that this high sounding word describes something that exists only in the imagination of some person. However. the books are full of it and Jacques Loeb. M. D. PH. D and S. C. D. (member of the Rockefeller Institute) in his book on the Physico-Chemical or Mechanistic theory of life tries to show that this fact is the most significant proof that life is merely a chemical force. Here is what he has to say on the subject: "The positively heliotropic animals which go instinctively to a source of light have in their eyes photo sensitive substances, which undergo chemical alterations by light. The products formed in this process influence the contraction of the muscles mostly indirectly through the nervous system. In a series of experiments I have shown that the heliotropical reactions of animals are identical with the heliotropical actions of plants. In plants only the more refrangible rays from green to blue have these heliotropical effects, while the red and yellow rays are little or less effective, and the same is true for the heliotropical reaction of animals."

Now here follow some of his experiments in the following words: "Some experiments on winged plant lice may serve as an introduction, etc. * * * In order to obtain the material, potted rose bushes infected with plant lice are brought into a room and placed in front of a closed window. If the plants are allowed to dry out, the aphides (plant lice), previously wingless, change into winged insects. After this metamorphosis the animals leave the plants, fly to the window and there creep upward on the glass." Then he makes the further remark: "It can be demonstrated in these animals that the direction of their progressive movements is just as unequivocally governed by the source of light as the direction of the movements of the planets is determined by the force of gravity."

Certainly Mr. Loeb must know that the cells who build these aphides or plant lice are in every way similar and identical to the cells that build him or Mr. Rockefeller. They are of exactly the same size and general appearance. I think the cells that build the plant lice show an extraordinary amount of skill and intelligence. As long as the rose bush was living, they were satisfied and contented because they had enough to both eat and drink. When Mr. Loeb dried up the rose bush and deprived them of food and drink, they had to get out of there and go some place where they could procure those things necessary to their life. They are compelled to do this just as Mr. Loeb would have to do under similar circumstances or starve to death. The plant lice could clearly see that there was nothing to eat and drink within any reasonable distance of where they were stranded. It looked to them like a "Long way to Tipperary." The only method by which they could save themselves from starvation was to prepare temporary wings and fly, which they did. Now where or in what direction should they fly? Anyone with sense would go straight to the place which would look like an opening in the prison or house in which they were enclosed. The light in the window would be the only thing that would direct them to a way to get out and find another green plant which is their only source of food and drink. What more could Mr. Loeb have done if he had been in their place? Endowed with the intelligence that Mr. Loeb and Mr. Rockefeller are now supposed to have, what more could they have done to save themselves from starvation? The intelligence, skill and knowledge possessed by the builders of the plant lice to be able to build wings with which to carry themselves away, when necessary, in an emergency like that, is in my opinion wonderful. It is far superior to ours.

Here is another illustration that Mr. Loeb gives of heliotropism: "If small crustaceans of a fresh water pond or lake are taken with a plankton net at noontime and placed in an aquarium which is illuminated from one side only, it is found that those animals move about in the vessel pretty much at random and distribute themselves irregularly. Some seem to go more towards the source of light, others in the opposite direction, and the majority, perhaps, pay no attention to the light. This condition changes instantly if we add to the water some acid. If the correct amount is added, all the individuals become actively positively heliotropic and move in as straight a line as the imperfection of the swimming movements permits towards the source of light and remain there closely crowded together on the illuminated side of the vessel. How does the acid produce this result? We will assume that it acts as a sensitizer." Mr. Loeb inquires, how does the acid produce this result and suggests that the acid probably is a sensitizer. We are forced to ask the question, that if a man were in the water in place of these animals and someone should put enough acid into the water so as to burn his skin and body, if he would not become just as heliotropic as these crustaceans and if he would not make a scramble to get out precisely in the same manner. Being confined in a dark cave like the aquarium with acid burning his body, his only thought would be to escape from the place. The only thing to guide his actions to effect an escape from this dark cave would be the light. It would be the only thing that could lead him or indicate to him an opening to the outside world. These cells or animals called crustaceans do nothing different from what you or I would do under the same circumstances. Then he makes the following comment: "In certain animals, for instance in daphnia and in certain marine copepods, a decrease in temperature also increases the tendency to positive heliotropism. If the mere addition of acid is not sufficient to make daphnia positively heliotropic, this may often be accomplished by simultaneously reducing the temperature." In other words, he might as well say that if you cannot start them moving by burning them with acid you can do so by freezing or making it uncomfortably cold for them. Why should they not try to escape from a place where they are freezing or where their skin is being irritated and eaten by acid?

I will quote you one more of his illustrations, which he thinks is very significant of the fact that some animals become at times possessed by a property that he calls heliotropism. I think it is the most absurd illustration of them all. Here is what he has to say about the young beetle, who comes out of the ground in the spring hungry and is very much in need of something to eat: "This change in the heliotropic sensitiveness produced by certain metabolic products in the animal body is of great biological significance. I pointed out in former papers that it serves to save the lives of the above mentioned young larvae of chrysorrhoea. When the young larvae are awakened from their winter sleep by the sunshine of the Spring, they are positively heliotropic. Their positive heliotropism leaves them no freedom of movement but forces them to creep straight upward to the top of a tree or branch. Here they find the first buds. In this way their heliotropism guides them to their food. Should they now remain positively heliotropic they would be held fast on the ends of the twigs and would starve to death.

but we have already mentioned that after having eaten, they once more lose their positive heliotropism, they can now creep downward until they reach a new leaf, the odor or tactile stimulus of which stops the progressive movements of the machine and sets their eating activity again in motion."

I want the reader to stop a moment here and consider the absurdity of this statement. The actions described by Mr. Loeb of this beetle are those of any intelligent being under the same circumstances. When the beetle comes out of the ground, he is hungry; he knows where to go to procure his food. Without having had any previous training or information as to where to find something to eat, he goes right after it. After having eaten one bud, he goes after the next one. This knowledge of where to go to find something to eat without having had any information on the subject, other scientists call instinct, but Mr. Loeb calls it heliotropism. He goes so far as to state that but for this heliotropism, the beetle would not know where to find his food and would starve to death. Why should the beetle lose his heliotropic property by eating a bud? Mr. Loeb states that this heliotropism would pin him to the top of the tree, but for the fact that it is removed by the eating of the bud. Why should the actions of the beetle, any more than those of Mr. Loeb's, be those of heliotropism when going in search of food? You may ask, how can the young beetle know where to go without previous experience or information as to where to find his food? That will be fully and clearly explained under the chapter on heredity and instinctive action. The reader must remember that the cells who build the young beetle also direct its action and course, just as we direct the action and course of a boat and other vehicles. The experience and knowledge are in the cell, the builder of the beetle, who was but last year the director of the parent, and well remembers where it got its food. There are other tropisms besides heliotropism which are used to describe the property of organisms being attracted by light. There is geotropism to show that they are attracted by the gravity of the earth and galvano tropism to denote that they are affected by an electric current. These words are used in demonstrating that life is merely a chemical action affected by light, electricity and gravity in the same manner as dead matter and chemical substances.

So far as we can see and understand the question at this time, life seems to be an intelligent being we call cell, moulding and directing matter for its own use and purpose. I am not going to say just what life and intelligence in the cell really are. I only want to get matters lined up right in reference to those things we actually know.

The cells in the brain of man are capable of directing his actions so as to effect the wonderful progress and inventions that have taken place. In building and maintaining the body, they act precisely in the same manner as we do. The cells which make up the lining of the stomach for instance, reach out their hands and select this and that from the mixed mass we have placed therein. They can always discriminate, for instance, between bits of fat and particles of coal, absorbing the former and leaving the latter. The surface cells pick out what is needed and hand it over to the other cells, who carry it to the place where it is wanted. Their acts show judgment and discretion, the same as our own.

There has been a vast number of experiments in late years and my opinion is based on these experiments and the conclusions that have been reached from them. Nearly all who are occupied in this work have been specializing.

Each one in his particular subject has been so completely occupied with his part of the whole that he has not taken time to summarize or comprehend life as a whole. It makes me think about the story of the five blind men who went to see what the elephant was like. One put his hands on the elephant's side; one put his arm around his leg; one got hold of his trunk; one took hold of his tail. As they were all blind and did not take time to feel of the whole animal, no one of them had any really correct idea of what the elephant was like. The one who got him by the leg thought he was like a tree, and the one who got him by the trunk thought he was like a snake, and the one that got him by the tail thought he was like a rope. It all depended on what place they had felt of him, and as their experiences were all different they never agreed as to what the elephant was like and never could; while a man who could see and had seen the whole elephant, could clearly hear from their discussion that they were all wrong. It appears to be similar with the scientists today. Each one feeling only of a certain spot is not competent to tell what life as a whole really is.

There are certain things that we all can see, that the microscope has revealed to us as clearly and as certainly as the food we eat and the clothes we wear. That is to say, it is undisputed that these microscopic animals exist. We find them as living, single, separate, animals or beings in both salt and fresh water. Every drop of water taken from ocean or pond contains one or more of them. It is undisputed that these beings build from the material at hand all the structures we see and know as plants and animals. Upon this base of facts admitted, we can and should be able to agree as to the cause of development in life, inheritance and instinctive action. Beyond these facts, we do not yet know the inner life of the cell itself.

We must stop now and then before the unknown. The more we become acquainted with this wonderful being we call cell, the more we see our ignorance of the depths of his real existence. Only a few years ago we knew very little about the life of the cell. Today we understand him a little better. Every day questions are put to the cell and some of the replies, though vague, reveal an unexpected purpose and intelligence in his actions.

I have stated enough to indicate the general purpose of this book and what I intend to prove by it, and shall now begin the argument by first considering what life really is as nearly as we can determine from all experiments made along that line up to the present time.

CHAPTER II.

WHAT IS LIFE?

Life began sometime in the past history of the earth or else with the beginning of time. If it is a property of matter and came to exist far back in the Paleozoic ages when the world was young, conditions must have first become suitable for it to appear. It did not appear as a cell or bacterium but as one of those primordial cells or ultramicroscopic beings, which we now find organized into the perfect animal we call cell. This primordial being must have existed for ages as a single separate individual before it began its social life in the cell It did not then understand how to transform solar energy into chemical energy and thereby make the food now used by us and the cells.

This primordial cell without doubt exists today in a single state. There must exist at the present time a whole world of living creatures which have never been seen by the microscope, of all sizes from the single separate primordial cell up to those cells we call bacteria, protozoa and the plant and animal building cells. We can never find out what life really is until we can invent a method by which we can study and understand the inner life of the primordial being which builds the cells. Up to the present time we can see that life exists only in this animal we call cell. A cell can come only from another cell. In our question as to what life really is, we can only pursue our investigation by demonstrating what life is not. We can show, I believe, that life is not any of the forces of nature with which we are now acquainted, except the force that we call intelligence. Considering the fact that the cell is the builder of all things we see on this planet that have what we call life, or exhibit that phenomenon we can do nothing except to go into a detailed investigation of the actions of that individual as compared with the actions of matter. We have seen from our physiology that the cell is either a one cell organism living a single separate life for himself only, or a colony of millions or more cells living together and all working together for the good of the whole community, as we find the case to be in a plant or animal. For that reason they have been divided into unicellular and multicellular. In this discussion of what life really is, we can do no more than compare the actions of these microscopic animals with the actions of matter, as they are affected by the forces of nature, known to us as gravitation, electricity, chemical force or affinity, light, heat, cold, wind, water, etc. I think we shall find that these forces will, that is to say, they will affect the cell in the same manner that they would affect us if we were similarly situated.

We shall try to see if possible if the phenomenon of life is any one, some or all of these forces. We shall begin with the force known as light. Light is the waves of a medium which occupies all space in the universe and is known as ether. Light is known as ether waves. It has been demonstrated lately that these waves are different, just as are the waves of the sea, and according to their size they produce different effects. Some produce heat, some light, and some electric effects. Ether waves may be polarized, reflected, absorbed and refracted. We do not have any experimental proof as to just how these things take place, so we do not know, but there seems to be no doubt that light, radiant heat and electric waves are all of the same nature. The ether waves are sometimes spoken of as visible or invisible light. The vibration of the ether set in motion by the sun is heat when felt by the hand and light when felt by the eye. Heat is understood to be and is a disturbance of the molecules of matter. The more violent the disturbance the more heat. The cells that build plants understand how to direct and use the heat of the sun. As the vibrations of the ether stir up the activity of the atoms of matter, the cell knows how to pile them up in large molecules, which are called colloids. These are merely building material and contain no life, such as proteids, fats, starches, sugar and other carbohydrates. These building materials known as colloids are such combinations of atoms as break up very easily and when released can be again directed to take such form and substance as may be required by cells in any particular place in the body. As the atoms change from one substance into another, they jar the surrounding bodies and the atmosphere in which they come in contact, and this produces what we call heat. This atomic motion and vibration is also called chemical energy. Now as it is a fact that only a part of the cells understand or in other words have the power to make their own food and building material from the raw material of earth and air by chemical action from the heat of the sun, it is clear that chemical action is not life nor the cause of life.

Possibly heat is not very generally understood and the following is a good short description of heat by Garret P. Serviss: "Heat is a violent agitation of the infinitesimal particles or molecules of which all matter is composed, hence there can be no heat in a vacuum where there are no particles of matter to be agitated or in the ether which is a medium that does not obey the ordinary laws of matter.

But on the other hand the vibrations that give rise to heat when they encounter material bodies must exist in the ether because it is the ether which carries them, and since ether is not excluded like matter from what we call a vacuum, it follows that these vibrations may exist in the vacuum, in fact, the only thing that a vacuum contains is ether. No doubt these vibrations conveyed from the sun by the ether may produce many other effects unperceived or unknown to us because we have no special nerves or organs of sensation suitable to their perception. The ether you will observe is a very mysterious thing and the discovery of its existence is one of the greatest triumphs of human intelligence. It seems to possess some of the properties of matter and yet it defies most of the laws of matter as we know them. To study all the phenomena of heat would require the devotion of an entire life time. Have you ever reflected upon the reason why heat can turn iron into a liquid and water into steam? Most persons seeing these things done or knowing they are done every day, think no more about it. It is not from among such persons that the great leaders of human advancement make their appearance. But to return to the answer to our question—iron is melted and water is vaporized by that very shaking or agitation of their constituent particles of which we have been speaking. The molecules or particles of a solid or a liquid are held together by the mutual attraction, not the attraction of gravitation but another sort of attraction called cohesion. The distance over which this kind of an attraction acts is very small. Each particle draws upon its immediate surrounding particles and they in turn upon others and thus the whole mass of a solid body or of a portion of liquid is held together.

"In solids the force of cohesion is so great that the particles are held in a rigid form. In liquids it is relatively so weak that the particles may slide about over one another, and in a gas or vapor there is no cohesion. Now when a solid is heated its particles are set into extraordinary vibration and if the heating is carried to a sufficient degree, the force of their cohesion will be so far weakened that they begin to slide over one another and the solid becomes a liquid. If the heating is carried still further, the particles will be so shaken that they lose their cohesive grip entirely, and the liquid expands into a vapor.

"As you sit in front of your winter fire and see the black coal or hard wood molecularly shaken asunder in the jaws of heat until part of it ascends in gases and part falls in ashes while the agitation sets up new waves of heat in the surrounding air and ether, you may if you will, become a philosopher and contribute your own little share to the thinking which drives the world."

This is a very good description of heat and particularly of the ether vibrations we receive from the sun. Mr. Serviss is of the same opinion as myself, that if we possessed organs of sensation suitable to perceive, we would be able to understand very many other effects chemical and otherwise, which the waves of ether from the sun produce on matter. No doubt the cell is in possession of the different organs of sensation suitable to perceive the effects that the solar heat or ether waves have on matter and is thereby able to direct the atoms to affect the large molecules in which he is able to store his building material and energy.

It is a hard and tedious work to forge and mold the molecules of the raw material of earth, air and water into food and building material, so a large number of cells have become parasites and live upon the work and energy of other cells. From this fact that some cells are able to use the heat from the sun's rays, to build and forge the material they require, and others not, it is clear that life is not caused by the heat of the sun but that the heat is used in the same manner and for a purpose, as it is used by intelligent man.

Coming back again to the question as to whether the light has a power to attract plants and animals, I believe it has, and that plants and animals act in the same manner as we do and for a purpose. They go towards the light for several reasons but mainly to get out of the place in which they are confined. Plant cells build their structures towards and into the light because they must use the rays of light or the waves of ether to effect the chemical action necessary to manufacture their food and the building material with which they build those structures we call plants. In Mr. Loeb's book "On the Mechanistic Conception of Life" he makes this statement: "At the present day nobody seriously questions the statement that the action of light upon organisms is primarily one of a chemical character."

It appears perfectly clear to me that the organisms or cells are themselves not in the least affected by the action of light any more than man except in this way, that if they are deprived of the sun's heat and light, they are deprived of a force with which they are able to manufacture those things that they must have to eat, and also those materials with which they build their structures. Sunlight seems to contain a peculiar combination of energy containing both heat and electricity, which effect chemical changes in the raw material they gather from earth, air and water and produce those particular things they must have for their existence. After Mr. Loeb has stated that light has certain effects upon the organism he makes the further statement: "While this chemical action is of the utmost importance for organisms the nutrition of which depends upon the action of chlorophyll, it becomes of less importance for organisms devoid of chlorophyll." The word chlorophyll is intended to mean a power in plant cells to make starch for food from the raw material of earth, air and water by the assistance of sunlight. You will notice from this statement that those cells who do not understand how to make starch by the aid of sunlight, are not attracted by sunlight in the manner that others are.

Matter at the present time is classified into about 80 elements. These elements can be again broken up into molecules, molecules into atoms, and the atoms into electrons. All things in this world seem to be in motion. There is no real rest anywhere. It is always only apparent or relative. Heat and light themselves, which constantly change, are merely forms of motion. One writer states: "In the eternal play of cosmic bodies, countless suns and planets rush hither and thither in infinite space. In every chemical composition and decomposition the atoms or smallest particles of matter are in motion and so are the molecules they compose."

We will find later upon further investigation that some cells carry with them a chemical laboratory where they are able to direct and forge the atoms into the larger molecules of matter to suit their needs.

Charles G. Gibson in his "Scientific Ideas of Today," states: "We picture the elementary atoms grouping themselves into little congregations called molecules. We picture the atoms grabbing hold of one another and we find that the different atoms have different grabbing powers. For instance, when we combine hydrogen and oxygen together, each oxygen atom is able to grab two hydrogen atoms to itself. Therefore, when we break up water into its constituent gases by means of an electric current, we find that we get twice the volume of hydrogen that we get of oxygen. The co-partnery agreement of the combination, known as 'water,' reads that there shall be two members of the hydrogen family, and one only of the oxygen family in the combination.

"In our common table salt there is a very simple copartnery, one atom of sodium combining with one atom of chlorine. Then again a single atom of gold will grab three atoms of chlorine to form the chloride of gold used in toning photographs. We see the elementary atoms with their electrical charges combining together and thus forming the neutral molecule, but even these molecules are far below the range of the most powerful microscope. We think of the invisible microbe again and try to realize that it contains millions upon millions of individual particles or molecules, each of which contain several atoms. We therefore picture a piece of solid iron as being entirely composed of invisible atoms of iron. If we handle a piece of solid iron it is very apparent that the invisible particles of which it is composed must have a powerful grip on one another. To this force which binds the molecules together, we have given the descriptive title of cohesion from the Latin word cohaereo, meaning I stick.

"Our thoughts naturally turn to the electrons, which along with the ether, are the most fundamental things of which we have any knowledge. It is evident that a very heavy burden falls upon those tiny charges of negative electricity. We have seen that they are the stuff that atoms are made of. Electric current and electric discharges are simply these tiny electrons in motion. We have also seen how the movements of electrons give rise to magnetic fields, electric waves, heat and visible light and every variety of ether waves.

"In the blazing sun we picture electrons revolving around myriads of atoms of matter at a great variety of speed. Why electrons go round some kinds of atoms faster than they do around others we shall see later. These electrons in the far distant sun are producing a great variety of different wave lengths in the ether. We see some atoms giving up one or more detachable electrons, which are accepted by other atoms producing a disturbance in their electric balance and causing the atoms to attract one another and become chemically united. In this way we account for the production of all the variety of all the compound substances known."

This statement by Mr. Gibson will give the reader a general idea of how matter seems to behave to the mind of the scientific man today. It is sufficient to give the reader, who is not familiar with these things, some idea of the nature of matter, at least so far as will be necessary to a reasonable understanding for discussion of the question at hand. The reader will see that so far the modes of energy exhibited by force and matter in the universe are not different from those in nature with which we have been heretofore familiar. The atoms and molecules of matter always follow fixed laws. We shall find that the actions of the cell are not those of fixed laws but contrary to them. He is a master of matter.

Does light affect the cells chemically or do they merely act as we would under similar circumstances? Mr. Loeb gave the following illustration to prove that certain cells are chemically affected by light: "When we observe a dense mass of copepods collected from a fresh water pond, we notice that some have a tendency to go to the light while others go in opposite directions and many, if not the majority, are indifferent to light. It is an easy matter to make the negatively heliotropic or the indifferent (copepods) almost instantly positively heliotropic by adding a small but definite amount of carbon dioxide in the form of carbonated water to the water in which the animals are contained. If the animals are contained in 50 C. C. of water, it suffices to add from 3 to 6 C. C. of carbon in water to make all the copepods energetically positively heliotropic. This heliotropism lasts about half an hour, probably until the carbon dioxide has again diffused into the air. Similar results may be obtained with any other acids."

I would like to ask what any man would do under similar circumstances if he was in a room and someone filled it with a poisonous or irritating gas? Would he not also at once become "energetically heliotropic" and attempt to escape from the room? Would he not also lose his heliotropism as soon as the noxious gases had passed out of the place where he was confined? I think he would in precisely the same manner as these cells, which are called copepods. The cell is an animal that feels, feeds and performs all the functions of life in every manner similar to that of Mr. Loeb. The advocates of the chemical theory of life base their strongest proof on this point of heliotropism, which is entirely without foundation. I do not think it is necessary to go into any further detailed discussion of the other tropisms like galvano-tropism and geo-tropism. The first is the idea that some cells are affected by electricity, the other that they are affected by the force of gravity. While it may be said they are affected, still they are affected in no wise different from ourselves. They will take advantage of the electric force and turn it to their own purpose similar to ourselves. For instance, the electric eel takes advantage of it and uses it in his business as a help in giving battle to his enemies, shocking and stinging them, thereby more easily capturing animals for food.

In reference to gravity it affects them in the same manner as it does a human being. In order to be able to get up into the sunlight, the cell will defy the law of gravitation and build his structure straight upward. The cell is an individual that is in no manner and in no way forced and pushed around by physical and chemical forces regardless of any will or opinion he may have in the matter himself. The physical forces are not able to cause him to act in any different manner from what they cause man to act.

There seem to be two sides only to the question, after looking over the enormous mass of philosophy and ideas written in the past in regard to the cause of life. One side claims that life came to exist sometime in the past history of the earth, out of the elements of the earth, and that it is only a chemical and mechanical phenomenon. The other side claims that a mind is back of matter, either separate from it or in the matter itself. Some call it vital force and some call it the Divine Will in the world or the universe. The plain reason for this difference of opinion arises from the fact that structures in nature show conclusively that intelligence of a high order or at least similar to that of man is the cause that forms and molds the different structures of life. We shall not in this chapter stop to discuss what intelligence is, but will leave that to a separate chapter.

It might be proper, however, at this time to give a short illustration of what is considered to be an intelligent act and one that is not intelligent and purely mechanical. I think the simplest illustration is the following given by

Prof. Wm. James: "The pursuance of future ends and choice of means for their attainment are thus the mark and criterion of the presence of mentality in a phenomena. We will use this test to discriminate between an intelligent act and a mechanical performance. We impute no mentality to sticks or stones because they never seem to move for the sake of anything, but always when pushed and then indifferently with no sign of choice, so we call them senseless."

Then Mr. James gives this illustration: "If some iron filings be sprinkled on a table and a magnet be brought near them, they will fly through the air for a certain distance and stick to its surface. Let a card cover the poles of the magnet and the filings will press forever against its surface without its ever occurring to them to pass around its sides and thus come in contact with the object of their love. Blow bubbles through a tube into the bottom of a pail of water; they will rise to the surface and mingle with the air. Their actions may be poetically interpreted as due to a longing to recombine with the mother atmosphere above the surface, but if you invert a jar over the pail they will rise and remain lodged beneath its bottom shut in from the outer air, although a slight deflection from their course at the outset or a redescent towards the rim of the jar when they found their upward course impeded would easily have set them free. If we now pass from such action as these to those of living things, we notice a striking difference. Romeo wants Juliet as the filings want the magnet but Romeo and Juliet, if a wall be built between them, do not remain idiotically pressing their faces against its opposite sides like the magnet and the filings with the card. Romeo goes over the wall or otherwise and touches Juliet's lips directly. With the filings the path is fixed. Whether it reaches the end depends on accident. With the lover, it is the end which is fixed. The path may be modified indefinitely. Suppose a living frog in the position in which we place our bubbles of air, viz., at the bottom of a jar of water—the want of air will soon make him also long to join the mother atmosphere, but if a jar of water be inverted over it he will not like the bubbles, perpetually press his nose against its unyielding roof, but will restlessly explore the neighborhood until he has discovered a path around its brim to the goal of his desire."

These illustrations are good and show clearly the difference between an intelligent and a mechanical act. I do not believe it is possible to show any intelligence in a purely chemical and mechanical act, as it never acts with a purpose. The cell has been active for millions of years in building structures like plants and animals. In the petrified forests of Arizona we see that trees were constructed in precisely the same shape and of the same size as they are today. We are satisfied that those trees were built by cells over two million years ago. After millions of years of practise and experience in building these stationary habitations for themselves like trees and plants, and movable structures like animals, it is reasonable that they should show a very high degree of skill and intelligence in this line of business. By the aid of sunlight they are able to handle and direct matter, molecules, atoms and electrons with the same dexterity that our best builders handle brick, mortar and cement.

The cells of our brain which do our thinking are not different from any of the cells of our body nor are they any different from the cells that live a single separate life in the water, nor from those cells that build plants. The slight difference in the general appearance of the cells of our body arises from the fact that they are occupied with different kinds of work, just as a preacher looks different from a blacksmith,—but otherwise they are just alike in general construction. All the different kinds of cells such as muscle cells, blood cells and bone cells came originally from one single cell. From this fact, it seems to me that no reason can be given why one cell is not as intelligent as the other.

Our intelligence is altogether based on our faculty of memory. Without it we could keep no record of our experiences. Without experiences to refer to we could form no judgments. From the experiences stored up in our memory we form judgments and ideas which guide our actions. Based upon this faculty of memory, intelligence is possible and from it all intelligent acts proceed. This power of the cells in our brain to receive and store away information and experience we shall see later is not only a property of the brain cells but of all the cells of the body. Memory is a peculiar thing. In old age we remember the things of childhood the longest. Sickness weakens the memory. Repeating an experience strengthens it. Drugs, fevers and excitement bring back to memory things long ago forgotten. However, upon this power of memory possessed by the cells all intelligence is based. Upon this power, we shall see, inheritance, instinct and reflex action are based.

Some go as far as to claim that matter acts as if possessed of intelligence and a will. For instance, if we drop a crystal of salt in water, the salt will disappear, the atoms of salt will move around in the water in perfect freedom. If the water is evaporated the atoms of salt will again come together as if they possessed a will to do this. You will notice, however, that this is a will that follows a fixed law. We can say the same of a falling stone. It always tries to go back to the earth. We find the same thing in chemical attraction and affinity, and some have gone so far as to call this "unconscious sensation," whatever may be understood by that expression. The word sensation is such a general term and subject to so many meanings, that it is hard to tell just what ideas they intend to convey. They have gone so far as to say that the actions of the atom in moving towards and joining other atoms to form other combinations of matter indicate a will and a soul in the atom. They claim that the actions of the atoms indicate a feeling of pleasure in getting together, and one of displeasure in being separated from other atoms. It seems clear to me that all these actions, sensations or motions of matter that we have so far been able to discover show actions under a fixed law that we find everywhere in the universe; while the intelligent action that we find in all organic beings we call alive, never moves under a fixed law, but always towards a fixed purpose regardless of those fixed laws of nature. I will admit that matter and force go together but some go further and claim that matter has both force and sensation. That is true if you wish to call action sensation. However, call it what you may, it is not intelligence.

We cannot deny the great difference between a living and a non-living body. The past history of the earth shows that life has left a trail of failures and successes, pain, carnage and extinction behind, in its struggle for existence. One writer, Mr. Burrows, states: "Man has taken his chances in the clash of blind matter and in the warfare of living forms. He has been the pet of no god; the favorite of no power on earth or in heaven. He is one of the fruits of the great cosmic tree and is subject to the same hazards and failures as the fruits of all other trees. The frosts may nip him in the bud; the storms beat him down; foes of earth or air prey upon him and hostile influences from all sides impede or mar him. The very forces that uphold him and furnish him with armory of tools and power will destroy him the moment he is off his guard. He is like the trainer of wild beasts who, at his peril, for one instant relaxes his mastery over them. Gravity, electricity, fire, flood, hurricane will crush or consume him. If his hand is unsteady or his wits tardy, nature has dealt with him as with all other forms of life. She has shown him no favor."

This is a very good description of the general condition that prevails in nature. This is a true description of the struggle for existence found everywhere in nature from the smallest bacteria to the largest cell, as well as among all plants and animals. We find the cell living singly in the ocean, in his separate struggle for existence, has made for himself a coat of armor from materials of different kinds, such as horn, lime and flint. Without sufficient intelligence to provide himself with these coverings, he would have been unable to battle with the crushing elements of the sea and to perpetuate his existence. It might be interesting to the reader to know what the great scientist, John Burroughs of the Rockefeller Institute, has to say in reference to what life really is. Here is part of an article written by Mr. Burroughs:

"Our studies of the past histories of the globe reveal the fact that life appeared upon a cooling planet when the temperature was suitable and when its basic elements, water and carbon dioxide were at hand. How it began, whether through insensible changes in the activities of inert matter lasting whole geologic ages or through sudden transformation at many points on the earth's surface, we can never know.

"But science can see no reason for believing that its beginning was other than natural. It was inevitable from the constitution of matter itself. Moreover, since the law of evolution seems of universal application and affords the key to more great problems than any other generalization of the human mind, one would say on primordial grounds that life is an evolution; that its genesis is to be sought in the inherent capacities and potentialities of matter itself. How else could it come? This is certainly the only natural road and it leads straight to the physico-chemical theory of the origin of life—the view held by an increasing number of biologists and bio-chemists of our day. It is the scientific view; no other view is possible to science as such. Science cannot go outside of matter and its laws for an explanation of any phenomena that appear in matter. It goes inside of matter instead and in its mysterious molecular attractions and repulsions in the whirl and dance of the atoms and electrons in their amazing potencies and activities, sees or seems to see the secret of the origin of life itself.

"To the scientist the earth is complete in itself. He can admit of no break or discontinuity anywhere. Threads of relation, visible and invisible, chemical, mechanical, electric, magnetic, solar, stellar, lunar, geologic and biologic—forming an intricate web of subtle forces and influences bind all things, living and dead, into cosmic unity.

"The disruptions and antagonisms which we fancy we see are only the result of our limited vision. Nature is not at war with itself. There is no room or need for miracles. There is no outside to the universe, because there are no bounds to matter or spirit. Science traces the chain of cause and effect everywhere and finds no break. It follows down animal life until it merges into vegetable, though it cannot put its finger or its microscope on the point where one ends and the other begins. It finds forms that partake of the characteristics of both. It is reasonable to expect that the vegetable merges into the mineral by the same insensible degree, and that the one becomes the other without any real discontinuity. The change, if we may call it such, probably takes place in the interior world of matter, among the primordial atoms where only the imagination can penetrate.

"Looked at in its relation to the whole, life appears like a transient phenomenon of matter. I will not say accidental; it seems inseparably bound up with cosmic processes, but I may say fugitive, superficial, circumscribed. Life comes and goes; it penetrates but a little way into the earth; it is confined to a certain range of temperature, beyond a certain degree of cold on the one hand it does not appear, and beyond a certain degree of heat on the other hand, it is cut off; without water or moisture it ceases, and without air it is not. It has evidently disappeared from the moon and probably from the inferior planets and it is doubtful if it has yet appeared on any of the superior planets, save Mars. Life comes to matter as the flowers come in the Spring when the time is ripe for it, and it disappears when the time is overripe. Man appears in due course and has his little day upon the earth but that day must as surely come to an end.

"Yet can we conceive of the end of physical order? The end of gravity? Or of cohesion? The air may disappear, the water may disappear, combustion may cease, but oxygen, hydrogen, nitrogen and carbon will continue somewhere."

This statement is very interesting in that it shows how life looks to a man in a chemical laboratory, who can by reason of his peculiar occupation see life only as a chemical force and action. He has had hold of the elephant's tail. To him the elephant is like a rope. You will notice that at no place does he even mention or consider the cell, who is the builder of all living things we see. He can see only the atoms, molecules, electrons, smashing around in space like a great cyclone, tearing through a city, and in this clash and crash of the blind forces, life started and continues today. In my opinion a chemist from his laboratory can form no opinion of the structure and life of the cell. A chemist deals with the crude forces of nature, with dead matter. The remains of a watch, a threshing machine or a human being after it had been burned into a gaseous form would not furnish a person with any information as to the nature, character, purpose or inner life of those structures.

The cell is an animal, very highly organized and specialized. Take the single cell called amoeba for instance. He has no machinery with which he can manufacture starch. He does, however, carry with him building material with which he can in an emergency save his life by covering himself with a coat of armor. Other cells carry with them a structure which is called chromatophore. With this instrument, these cells are able to manufacture starch from the crude substances of earth, air and water by the aid of sunlight. From these facts, it must appear evident to the reader that the cell is a very highly organized and specialized individual, and that to look at him from the point of view of being mere matter and force is the same as to compare the actions of a stone rolling down a hill with that of an automobile moving over a smooth pavement. One is compelled to move by reason of the force of gravitation while the other moves by virtue of the intellect that guides it. The structures of life, like plants and animals, are built from the materials taken from the earth, air, and water, just as are the structures man builds, like railroads and skyscrapers. If we were asked how it is possible for man to effect the construction of these railroads and buildings, we would say that it is by reason of the fact that he is an intelligent being.

The intelligence of man is the intelligence possessed by the cells in his brain. If man is intelligent and by virtue thereof is able to combine and arrange matter and force so as to effect structures such as houses and railroads, why is not the cell also intelligent when he is able to direct the forces of nature so as to effect the structures we see such as plants and animals. The cell is not compelled to act by reason of any chemical and mechanical force, any more than is man. He acts by reason of a will and judgment of his own. He is a separate living animal. I see no reason to impute intelligence to the activities of atoms and molecules any more than to bricks and stones. Bergson in his "Creative Evolution" seems to see in matter and life a creative energy. If we stood at a distance watching a skyscraper gradually grow into completeness, we would say there must be some creative energy back of it, pushing the construction and, if we could never get near enough to see the men and builders at work we could have no other idea of how that sky scraper came into existence except that it was caused by some creative energy.

How do we know that the cell does actually build all organic materials or living structure just as man builds his larger structures, like machines, houses, railroads, etc.? First—because we can now see him do it.

Second—because matter itself is such that it never could and never will produce anything living. Is it possible to demonstrate that all matter has this quality? I think it is. Wherever a tree is growing or sprouting, there the dead matter is being transformed into the living, but wherever a tree or animal is dead or decaying, it is gradually again turning into the organic matter. So it is with structures produced by man. While a house or city is occupied by the builders thereof, it is being maintained or growing, so to speak, but as soon as the people are destroyed, the structures gradually return to the ordinary matter of the earth.

We find the fossil remains of the structures of man, like old ruins of cities, just as we find fossil remains of the structures of the cell, like the bones of animals. Matter and force and chemical action act just the same in one place as in the other. They are the same and will act the same in your stomach as in the sun or any other place in the universe. We know that nourishment is transformed into the body tissues and through different channels it leaves the body in precisely the same quantity as it entered, partly unmodified and in other form. No atom of matter has been lost. Digestion is a chemical and mechanical process. The cells are compelled to comply with and take advantage of chemical forces, matter and energy, in precisely the same manner that man does. The cells must keep up a continual chemical laboratory and mix and decompose substances according to the general laws of chemical affinity. They must be expert in their work and so it is with man, he can do nothing unless he is intelligent and understands his work. The elements that make up the plant or animal are the same as those of a house or battleship. The cells of our body will handle the food furnished them in exactly the same manner as the raw material furnished to a factory. The coal, iron and other raw material are changed into the product intended or required.

The plant building cells have produced the material for the cells of our body in the same manner that the miners have produced the raw material for our factories. In the factory as in the body, the laws of matter and force must be complied with. Water will penetrate, flow and sink according to laws of gravity, in one place as well as in another. The knowledge of laws of matter and force in the universe is as necessary in changing crude material into the finished product in the factory as in our bodies. The laws of chemistry and mechanics must be observed by the people in the factory as well as by the cells in our body or in plants. The circulation of the blood is a purely mechanical act, like that of man pumping water. The cells that produce milk or digestive fluid must employ chemistry. The chemist will analyze the crude elements in a tree or animal and find that it contains the same elements as are found everywhere in the universe, and from these facts he will draw the conclusion that life itself is only a chemical action. It seems to me absurd to make such a statement. If you should crush a skyscraper or battleship with all its inhabitants and analyze its mangled mass of iron, mortar, wood, brick, and human beings, you could not possibly find anything but the ordinary elements of matter found in the universe; still it would be clearly absurd to say that the skyscraper or battleship had produced itself from the raw materials, or that the iron, brick and mortar had produced the skyscraper, or that the wood, metals and other material had produced the battleship. There are several things now made by man that were produced a few years ago only by the cell. For instance, free nitrogen, dextrose, several organic acids, perfumes, candles, Berlin blue, taurin, etc., were manufactured exclusively by the cells until recently, when man also discovered how to make them.

Chemists are discovering how to make these products of life made by the cell, and no doubt in the future will be able to make many more, as we have plenty of the raw material from which to make them. We must, however, remember that chemical force and action is the same everywhere, and that the cell must plan, mix and guide the actions of the chemical forces and matter just as man does in order to obtain and produce these artificial compounds found in life. If intellect is not there to guide and direct the matter and force of the universe, nothing will be produced. Heat influences the actions of the atoms and molecules of matter. Even chemical attraction gives way to heat, so that all bodies at sufficient temperature are decomposed into free atoms or elementary parts. In this way heat performs a work in so far as it separates masses from each other, consequently a certain amount of mechanical work is equivalent to a certain quantity of heat and vice versa. In chemical action a transformation of energy of one kind into another takes place. The mechanical energy of the atoms is converted into heat, which may again be used for the other forms of mechanical energy. This explains why heat is developed in a chemical process. Every chemical process can be called a combustion. In a violent disturbance of the atoms and molecules of matter, we have the common phenomena of fire and light. The fact is that heat is a source of stored or convertible energy. The only source of heat at the surface of the earth is the sun. The cell must obtain heat or energy somewhere with which to produce the atomic changes and molecular disturbances or chemical actions and changes desired. Without heat the cell could not produce the products required like fats, starches and sugars, any more than man could produce his works of art or products from the factory where chemical action or heat is required. The productions of the cell with which plants and animals are put together are products of art. They are products that the forces of nature cannot produce. The irons and metals can never build an engine nor can the stones and brick produce a house.

The blind forces of nature must be guided by an intellect toward a purpose. To build a machine the ore must be lifted from the mines, smelted into plates and bars. These again must be forged into the shapes and sizes to fit into every part of the machine. The cell will produce food and building material for future use in precisely the same manner as man will get the iron ore from the mines and melt it into plates for future use. The intellect of cell or man shows the same wisdom, foresight, and purpose. The laws of matter and motion, cause and effect which we have found are always the same in the universe, will prevent the ores from coming out of the mountain and building a steam engine. These same laws will also prevent the material forces from producing plants and animals. There must be a living being with an intellect to guide the matter and force towards a fixed purpose.

The heat required to melt iron and metals is usually very high, so the cell does not use iron, brass, copper, etc., but such material as carbon, hydrogen, sulphur, phosphorous, chlorin, potassium, sodium, magnesia, etc. By the aid of the heat obtained from the sun, the cell is able to mix and combine the atoms of these elements into such material as may be required for future use in building up and maintaining himself, plants or animals. The reduction of carbonic acid and water into carbohydrates is done through the assistance and direction of the cells and from the sun they get their power and energy. Left to itself, we know that the sun could produce nothing. Products of art must be resorted to just as man takes advantage of photograph cameras and lenses or burning mirrors. It is very clear that the cell also must use similar artificial means with which to accomplish the results required. It seems clear that the cells have invented, constructed and possess artificial devices with which they can gather and direct the heat or energy of the sun and thereby mold matter and direct the actions of the atoms of matter as they wish. At any rate we know that the sun may shine on carbonic acid and water until eternity without producing fats, carbohydrates and proteids. Without the intellect of the living being we call cell interfering, they would not be produced. The carbohydrates produced by the cells of plants are products of art and so are all the products of cells.

Knowing it to be a scientific fact that matter and force, gravity, chemical affinity, etc. will behave just the same everywhere in the universe, we know that the cell (whom we also know to be a living animal or being) must employ the same methods as man to effect his purpose and desires. Without intelligence man could not produce his products of art and it must be perfectly clear that the cell must employ the same means and be possessed of the same quality or power in order to be able to produce his products.

Years ago before we had produced a microscope powerful enough to see the cell, we could see particles of matter take their place and arrange themselves in order so as to become collectively plants or animals, whose parts bore a strict relation to the whole. From our knowledge of matter we knew that it could not act in that way towards a purpose unless it was guided by the intellect of some being similar to our own, no more so than brick and stone could take their place in the construction of a house without being guided by the intellect of some living being. The thinker and observer could then as he can today, see that matter was guided by a mind similar to his own, towards a purpose. There seemed to be an invisible spirit in charge, directing atoms and particles of matter, which also proved to be the fact.

A being that is invisible is called a spirit. The cell is now no longer the spiritual being which he was a few years ago, but is a material living being. We can now see this being in charge and that he is the cause of matter growing into a plant or animal. He is the cause in the same manner as man is the cause of bricks and stones growing into a house. The designers and builders of plants and animals were to us spiritual beings because we could not see them. In the same way the designers and builders of skyscrapers and battleships would be spiritual beings to us if we could not get near enough to the structures to see the builders. The builders of the cell are at this time passing from the spiritual condition into the material, because at this time we can just barely see them and not plainly enough to be able to say just what they are. We find the cell is no primitive organism. He is again a colony of primordial beings, which are, in fact, the real actors and workers.

The cell is an animal that has evolved naturally from the smaller primordial beings in the same manner that plants and animals have come to exist by natural evolution of the cell. It would seem that the cells who can make food and building material by the aid of heat from the sun, came into existence first. It is not likely that the heat from the sun was used by the cells until after the earth had cooled down to a certain temperature. It is likely that the heat used by the cells ages ago in the huge production of vegetable matter disclosed by the coal deposits was the heat from the earth and not the sun. It is generally understood that life cannot exist without sunlight, air, water and food, but it begins to appear now that life can exist almost indefinitely without any of these. The seeds from many plants, I know from my own knowledge, can live from five to fifteen years, and it is claimed that some can live a much longer time under conditions where they are not attacked by other cells or bacteria. This also goes to show that the cell must have fuel, air, heat and water when he is active producing living structures just as man must have when he is building and running factories, railroads, etc.

The microscope disclosed to us a new world. If they had told us years ago that a drop of water contained hundreds of living animals that eat, drink, fight, love and reproduce, they would have had a hard time making us believe it. The school boy or anyone else today is compelled to admit that it is a fact, because through a microscope he can see it with his own eye.

Now since a world of teeming animal life really existed, which was not known to us because we did not have eyes suitable or powerful enough with which to see it, does it not seem plausible that there must still exist a world of still more microscopic proportions, which we shall probably see some day? When that time comes the builder of the cell will no more be a spiritual being, but probably a material living animal or being as the cell is today. It is easy to see what the materials are which the plant cells work up into building material for their habitation or colonies, which we call plants and animals. We also know that they use the heat and energy of the sun as a power. The atmosphere contains oxygen, nitrogen, carbonic acid and ammonia. The soil contains silica, iron, lime, potash, phosphorous, sulphur and ammoniacal salts. The soil and atmosphere contain all the material which is found in animals and plants and in the same manner we can point out the raw material contained in a railroad or battleship, but the cells, the builders, must separate, join and place the material in exactly the right place and in correct proportions to effect the purpose desired. To do this it is just as necessary for the cell to be skilled and intelligent as it is for man. It has been repeatedly shown that the smallest cell, known as germ or bacterium, is constantly changing its habits and methods of life, and that any common, harmless species of germ may change its method of living and become a disease germ. Leahman and Neuman, the best authorities on this subject, state: "The division of bacteria into pathogenic and non-pathogenic, etc. as is still always done in textbooks, has failed absolutely. We can understand and know the pathogenic variety only if we study simultaneously the non-pathogenic from which the former have once originated and still always originate." They then go on and show that the different disease germs, such as typhoid, diphtheria, scarlet fever, etc., can be changed from one to the other by cultivation in different places and conditions. This is very significant to show how life in the microscopic world is a struggle for existence and how the cell in every place in life prepares and adapts itself to conditions. It is the same in the microscopic world as in our world, everything is in a state of evolution and change there, as well as elsewhere. Evolution itself shows that there is a struggle for a purpose. Evolution really proves intelligence because it means progress step by step. The cell will progress and build a tree or animal step by step in the same manner that man produces his structures.

In order to cause matter to evolve in a certain direction to produce a certain end or structure, and none other, the material must be directed and guided by intelligence, as we do not find property or matter anywhere with a tendency to develop towards any purpose. Desire and will of the cell must be back of matter to produce the organic or living things we see. Man, animals, plants and cells all show the same intellect in their places in life. How could it be otherwise when we consider the fact that man, animals and plants are also cells? If the products and industry of man arise from his desire and ideas, so must also the industry and products of the cell. Insects will build a nest of clay and other material, lay their eggs in it and provide food for the young, which they will never see. Insects will do all these things without having had any previous instructions. The philosopher and thinker would observe these marvelous actions and adaptations involving so many different ideas, and he would say, how is it possible for mere matter to pursue a purpose involving such complicated plans and combinations of ideas? He would say that the insect or matter of which it was composed was in charge of or guided by a Divine Will or invisible being or spirit. We find now that the thinker and observer was correct.

The insect was, in fact, built by, was in the charge of and guided by an intelligent living being we call cell. They were to us at that time invisible, so they were correctly called Spiritual Beings. We see battleships and submarines move about on the surface of the water like ducks. They signal each other in like manner, and from all appearances and from their actions an observer and thinker would be compelled to say, looking at them from a distance, that they were directed and guided by spirits or some other intelligence. However, if he could use field glasses or the telescope or could get nearer, so as to be able to see the beings in charge of the boats, he would discover the beings or animals in charge of the battleships and these spirits would change into material living beings and the mystery would be solved. Every part of a body or plant reveals its use—for instance, the lungs reveal their use. It is the same in the battleship or other structure produced by man. Every part of ship or machine is made for a certain purpose. Oxygen of the air is necessary to produce heat and power in the battleship, so chimneys are provided to draw the air into the furnace. The gills are machines to separate the oxygen from water, and lungs to separate it from air. When we see the parts we know for what purpose they have been made. When we see a part of a battleship or railroad, we know what it is for unless we are very ignorant like the savage. In like manner when we see a part of a plant or animal we know what it is made for and that it is made by the cell.

The intention and purpose reveals the same intellect in both cases. Just as man in possession of the same building material of metals, bricks, wood, etc. can construct buildings and machines very different from each other, so also can the cells with the material they possess in the elements, produce organs of plants and animals very different in external form and appearance. The savage or an ignorant man looking at a battleship or complicated machine, would have no comprehension how the ship or machine had come to exist or had been put together. So today, too, many people who have not studied life and the cause and composition of living structures, cannot form any conception of what a plant or animal really is.

The brain of the bee contains only a few cells and is invisible except when viewed by the microscope. Still these cells in the brain of the bee, who direct all his actions, will manage a very effective battle against intelligent man, as will appear from the following, from my daily paper:

"Oakdale, Cal.—Forty school children were held prisoners for half a day at Langworth school recently by a buzzing swarm of bees, which finally broke up the school for the day.

"The bees had lived in one corner of the roof for weeks, and had been undisturbed until some of the youngsters threw clods and dislodged the hive. The bees attacked their tormentors, who took refuge in the school house. Miss Ida Warford, the teacher, put her head out of the door to see the cause of the commotion and was stung on the nose. Hundreds of bees swarmed into the half opened door and the children sought refuge in the next room while the teacher and the older boys did battle with the bees with wet cloths and whatever weapons they had handy. They, too, were finally forced into the other room, and the entire school was made prisoner until some of the parents, alarmed at the absence of their children, came to the rescue. They were forced to flee, too, but finally came back armed with sulphur with which they routed the bees."

What possible difference can there be in the intelligence evidenced by these people and the bees, fighting each other. The bees will behave and bother no one if left alone but in defence of their home and colony, they will fight to the last. In what manner do the actions of man differ from those of the bee? The cells that make the bee have had to fight animals for ages, and they are equipped with a poisoned dagger of a deadly character. Man was originally made not to fight but to escape from enemies through the trees from branch to branch. For that reason, he is a very helpless victim in a fight with other animals, but since his discovery of the club he has been king in the animal world, and has since been progressing very rapidly. While he was fighting only a few years ago with clubs, spears, bows and arrows, etc., he is now fighting with liquid fire. The bees first conquered the teacher and the school children and then chased the farmers home. The farmers came back with reinforcements and fought the bees with fire and poisonous gases. The bees fight with poisoned daggers and man fights with deadly gases. Both sides fight with poison in defence of their children. In what manner do they differ in their actions as intelligent beings? The cells building the bee must make the dagger and prepare the poison for a purpose, just as man does in his actions of foresight and preparedness. The cells in the head of the bee, as well as in man, are the parties who direct all the actions of this battle, and I fail to see any difference in their intellectual capacity.

Not long ago we had no microscope enabling us to see the builders of a plant or animal. At that time we were compelled to say that they simply grew but now we have seen the builder, and the next question is,—how are those builders able to build these structures? In order to be able to discover the real cause, it will be necessary to first inquire into the inner life of these builders, how they live, eat and multiply, how and where they get their ideas and building material. That would be the only sensible way to find out how the sky scraper was constructed. We have penetrated far enough into the mysteries of matter, molecule, atom, and electron. We can at this time clearly see that matter and force in the universe is everywhere the same; they always follow fixed laws and cannot be destroyed. The form and nature of the energy can be changed from one to the other; for instance, in the dynamo we see mechanical energy transformed into electrical energy, which in turn may be transformed into heat energy in the electric furnace; or the electric energy may be transmitted to a distance and be converted once more into mechanical motion by means of the electric motor. The kinetic energy of a waterfall may be transformed into mechanical energy by means of the old-fashioned mill wheel, and we might go on in this manner considering one series of transformations after another. The force of a living being is different from these natural forces in this, that it directs and guides the natural forces to its own purposes.

Not many years ago we did not know that cells produced the trees, plants and animals we see,—we did not know that those builders existed. When I first became interested in nature and biology, we were told by the scientists that the cause of plant growth was endosmosis and crystallization. We know now that the plants and animals that we see are structures produced by smaller animals we call cells. By reason of a more powerful eye made partly by ourselves and partly by the cell, we can see these individuals. The cell makes our eyes as nearly suitable for every day use as he thinks best, but for a further special use like seeing smaller things like cells, or for distant objects like planets, we add to or change the arrangement of the lenses of the eye to effect the purpose desired. It requires just as much intelligence to build the eye in the first place as to change it for any special or particular purpose afterwards. The cell is not a mass of matter composed of electrons, atoms or molecules, but is a highly organized and specialized living being. We shall find that he is made up of possibly millions of still smaller living beings. It seems clear that the only method to solve the mystery of life is to more completely investigate the inner life of the cell and put all the questions to him.

It seems to me the only reason that so many ideas and differences of opinion about life have arisen is the fact that we have not been able to look upon the cell as the primary source of life and intelligence. The mechanical side could see nothing but matter and force, the other side could see nothing but mind and intelligence back of it. The fact seems to be clear, that the same mind and intelligence is back of the production of plants and animals, in the same way and to the same extent that mind and intelligence is back of all those structures produced by man. In either case the mind and intelligence is and must be in the actual builder of these structures.

We know that it is not matter that is building plants and animals. It is the microscopic wonder we call the cell. We must remember that life came into existence millions of years ago and how it started we can only guess. It has had time to develop in so many millions of ways that we cannot hope to comprehend just how it came to be in the condition that we now find it. We all have a right to guess on that subject. We know that life is not heat because it can exist in a frozen condition for a long time; the seeds of wild plants live in the frozen north for years. Some experiments of late go to show that fish can be frozen for months in solid ice and live when thawed out, as will be seen from the following clipping from a Scientific Journal: "The feat of freezing live fish and reviving them several weeks or months later has been achieved by the Swiss scientist, M. Pictet. The scientist put twenty-eight live fish in a box that contained water rich in oxygen, in which several pieces of ice floated. The temperature of the water was then reduced slowly until it froze. At the end of about two months the cake was gradually thawed and the fish, it is said, were found alive. In such an experiment, the scientist reports, it is essential that the water be gradually frozen and that it shall have contained pieces of ice for from fifteen to eighteen hours before the whole mass is frozen. The process of thawing must also be slow.

"Through this process it is believed that Siberian sturgeon and Alaskan salmon can be exported alive to distant markets."

It is clear that life is not sunlight nor has sunlight produced life because the lower organisms or cells such as bacteria are disorganized and destroyed by sunlight. We also know that life is not electricity or magnetism because we see this force employed only by some animals and not by others as in the great electric eel of Africa where the electric organs weigh more than one-third of the entire fish. Life is not any of the forces of nature that we have yet discovered. The cell is the animal that contains life. We contain life because we contain the cells, in the same manner as a ship contains life because it contains people. The cell builds with the crude elements of matter and force in the universe in the same manner as man does. In answer then to the question, what is life,—we are compelled to say that it is the activities and products of the living animal or being we call cell.

CHAPTER III.

THE CELL.

We shall now investigate the inner life of the cell as far as we can go with the evidence at hand up to the present time.

I shall try not to deal with the speculative side of the subject any more than will be necessary to show up some of the problems yet to be solved. Some of the theories will be referred to in order to show where we are at, and which will point to further investigation of the facts. This chapter will attempt to discuss the actions of the cell in reference to his inner life, his methods, appearance and actions. No distinction will be made between animal building cells, single cells and plant building cells, as there is no difference in their primary method of living and reproduction.

The reader must understand that in this short chapter no complete history or description of cell life can be attempted.. All I can do is to briefly outline those facts about the cell that now seem to be admitted, so that the reader can see for himself what kind of an animal or being the cell appears to be. We shall consider the cell mainly under two heads: First—Its general appearance, structure and organs. Second—How they multiply and increase in numbers.

In my investigation of books discussing the life of the cell, I found Edmund B. Wilson, professor of zoology of

Columbia University, New York, about the best writer on this subject and shall quote him more than anyone else, to show the reader what we know about the cell at this time. I think he is very careful in his statements, and does not state any propositions as true unless admitted and backed up by very good authority. The following is a part of his introduction to the study of the life of the cell:

"Among the lowest forms at the base of the series are an immense number of microscopic plants and animals, familiar examples of which are the bacteria, diatoms, rhizopods and Infusoria, in which the entire body consists of a single cell, of the same general type as those which in the higher multicellular forms are associated to form one organic whole. Structurally, therefore, the multicellular body is in a certain sense comparable with a colony or aggregation of the lower one celled forms. This comparison is not less suggestive to the physiologist than to the morphologist. In the one celled forms all of the vital functions are performed by a single cell. In the multicellular forms, on the other hand, these functions are not equally performed by all the cells, but in varying degree distributed among them, the cells thus falling into physiological groups or tissues, each of which is especially devoted to the performance of a specific function. Thus arises the "physiological division of labor" through which alone the highest development of vital activity becomes possible and thus the cell becomes a unit not merely of structure but also of function. Each bodily function and even the life of the organism as a whole may thus in one sense be regarded as resultant, arising through the integration of a vast number of cell-activities; and it cannot be adequately investigated without the study of the individual cell activities that lie at its root."

This description is practically the same as the one I quoted earlier from our High School Physiology. You notice the same statement, that whether the structures of life are plant or animal they are based on the cell. Now the first thing the reader will want to know is this, has the cell special organs and how does it look? Fig. 3 is a

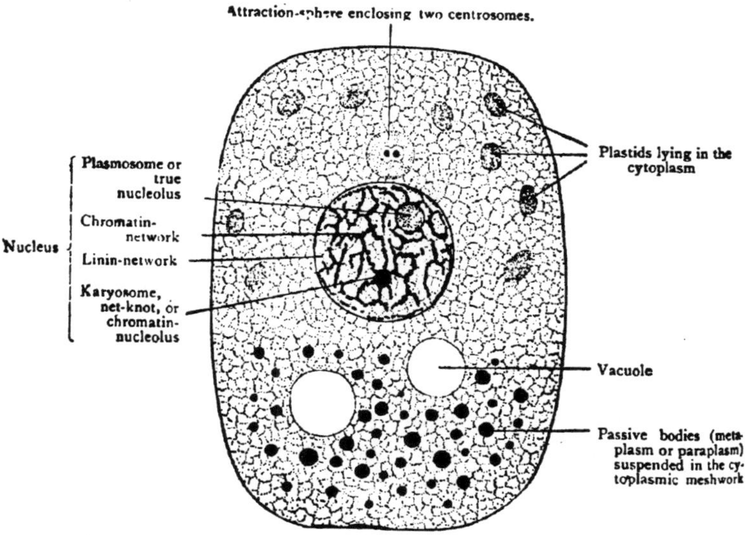

FIG. 3.—The Diagram of a cell. Its basis consists of a meshwork containing numerous minute granules *(microsomes)* and traversing a transparent ground-substance.—WILSON.

general sketch of this animal as it looks through a powerful microscope. The cell is a complete animal made up of still smaller individuals and organs just as a larger animal is. It has a head or directing center, which seems to direct the actions of the other parts. This directing center is called the centrosome. Then it has a bunch of sub-heads or skilled workers who are located in the middle of the body of the cell. These skilled workers have charge of all work in general. They seem to be the part of the cell which contains the power, knowledge and skill to perform the different kinds of work which the cell is required to do in order to exist. These specifically skilled workers located in the middle of the body of the cell are called the nucleus, and appear to be not one individual, but a very large colony of individuals. That this part of the cell called the nucleus is the part which has the power and knowledge of how to build the different structures in life, is shown by the fact that if this is destroyed, the cell cannot do any more work nor reproduce itself nor feed itself. In the same manner an animal is made helpless and is generally destroyed by the removal of its head. Besides the body, head, sub-head or skilled worker, it also has some sort of covering and a number of other special organs, not yet understood.

The cells are not all of the same size. Some are more highly organized than others and very likely contain a larger number of the primordial cells of which they are composed, and other special purpose cells. The smallest are the bacteria; then come the fungi and plant cells; the largest are the animal building cells and those similar to them who live separate lives in the water and do not build colonies like plants and animals. My school book on botany describes these smaller cells such as bacteria, plant cells and fungi in the following language:

"It would be hard to imagine a simpler plant, and the plant kingdom can be thought of as beginning with individuals consisting of one green cell and reproducing by division. This one cell, however, absorbs material, makes food, assimilates it, conducts respiration, etc., in fact, does all the work of living carried on by plants with roots, stems and leaves, although they may contain millions of cells. . . . .

"Bounding the cell there is a thin elastic cell-wall composed of a substance called cellulose. The cell-wall, therefore, constitutes a delicate sac, which contains the living substance. It is the substance that has formed the wall about itself in the same sense that a snail deposits the shell about its body. It is organized into various structures which are called organs of the cell. One of the most conspicuous organs is the nucleus, a comparatively compact and usually a spherical body and generally centrally placed within the cell.

"The fungi do not contain chlorophyll and this fact forms the sharpest contrast between them and the algae. The presence of the chlorophyll enables the algae to be independent of any other organism since they can manufacture their food out of carbon-dioxide and water. The absence of chlorophyll compels the fungi to be dependent upon other organisms for their food. This food is obtained in two general ways: either (1) directly from living plants and animals, or (2) from organic waste products or dead bodies. In case a living body is attacked, the attacking fungus is called a parasite and the plant or animal attacked, the host. In case the food is obtained in the other way, the fungus is called a saprophyte. For example—the rust that attacks wheat is a parasite and the wheat is the host; while the mold which often develops on stale bread is a saprophyte.

"Bacteria include the smallest known living forms. Even to distinguish ordinary bacteria, the highest powers of the microscope are necessary and to study them is too difficult for the untrained student. However, they are so very important to man on account of their useful and destructive operations that every student should have some information about them. Public attention has been drawn to them chiefly on account of the part they play in many infectious diseases, in which connection they are often referred to as 'microbes' or 'germs'."

I have now given a general description of the different classes of cells known as plant cells, animal cells, single cells and bacteria. However, all scientists agree that they are all of the same family; that they are all alike as far as inner structure of life is concerned; that they differ only in size, outside covering and appearance. They use different methods of obtaining their food and perpetuating their existence.

I wish to have this part perfectly clear to the reader, that all living things are either cells living singly and alone as separate individuals which we call single cells, like bacteria and others, or else a colony of cells numbering up into the billions, like plants, animals or trees, where the cells all work together for the benefit of all. As long as the tree or animal lives, they all live, but if the tree or animal dies, it is the cells in the tree or animal that die. By reason of the higher power microscope now made, it has been shown that the cell is made up of still smaller cells. These smaller units of life, which I would call primordial cells, have been described by various authors under a number of different names. The following are some of the names given to these hypothetical units of which the cell is supposed to be composed: Gemmules, pangens, plasomes, micellae, Plastidules, Biophores, bioplasts, somacules, idioplasms, idiosomes, biogens, microsomes, gemmae. This is only a partial list of the names and they all mean the same. Each name represents a different author and generally a different theory. While these theories are interesting, they are as yet only theories, as the units that make up the cell are too small to be clearly seen by any microscope yet invented.

It must be remembered that the cell organs that we are able to see like the centrosome, nucleus, vacuole, chromotophore and many others, are large crowds of these units located at different places in the cell body to perform special and different kinds of work. Some digest food, some effect respiration, some move the body, some manufacture food, etc. As these crowds are divided up in the cell to do different kinds of work, they look different. In this way the cell has organs with which to perform its different kinds of work, just as our bodies have organs to effect different kinds of work. In reference to this, Wilson states:

"Closely interrelated as the cell organs are, they have a remarkable degree of morphological independence. They assimilate food, grow and divide and perform their own characteristic actions, like co-existant but independent organisms of a lower grade than the cell, living together in colonial, or symbionic association. Yet we may still inquire whether the power of division shown by such protoplasm masses as plastids, chromosomes, centrosomes, nuclei, may not have its root in a like power residing in ultimate protoplasmic units of which they are made up. On the strength of these facts Boveri concluded that the chromosomes must be regarded as 'individuals' or elementary organisms that have an independent existence in the cell.

"The highest power of our present microscopes have not laid bare the ultimate organization of the cell. The cell might be composed of more elementary units ranking between the molecule and the cell.

"Whether the plastids arise solely by division or also by new formation, the foregoing observations on the plastids give a substantial basis for the hypothesis that protoplasm may be built of minute dividing bodies, which form its ultimate structural basis. . . . . . The cell is, in Burke's words, an elementary organism, which may by itself perform all the characteristic operations of life, as in the case with a unicellular organism. . . . Even when the cell is but a constituent unit of a higher grade or organization as in multicellular forms, it is no less truly an organism and in a measure leads an independent life, even though its functions be restricted and subordinated to the common life."

These statements by Prof. Wilson show that the cell, whether living his separate life in the water or other places, or acting as one of the units that make up the individual animal or plant, is an animal that has the power to perform all the functions of life. They also show clearly that all the different and separate special organs of the cell are crowds or colonies of still smaller cells or units that lead individual and separate lives within the body of the cell; that they feed and multiply within the cell in the same manner that the cell lives a separate individual life within the body of plants and animals. It shows that when the cell divides it is simply a division of a crowd of specifically skilled workers or beings and that when the cell multiplies by division each colony of specifically skilled workers in the centrosome and nucleus divides in two and each half then multiplies until it again reaches its original number and size when it is ready to divide again. The most wonderful thing is the centrosome or the directing center of the cell, which when the cell divides has charge of the work of looking after and seeing that the different crowds of specifically skilled workers are equally divided.

The centrosome seems to be the center of intelligence

and will power of the cell. In regard to this part of the cell Prof. Wilson makes the following statement: "From our present point of view, the centrosome possesses a peculiar interest as a cell organ which may be scarcely larger than a cytomicrosome, yet possesses specific physiological properties, assimilates, grows, divides and may persist from cell to cell without loss of identity. Nearly all observers of the centrosome have found it lying in the cytoplasm outside the nucleus, but apart from the protozoa there is at least one established case in which it lies within the nucleus, a fact that proves that its position is nonessential; that the centrosome is an active center rather than a passive body or one created by the aster formation is strongly indicated by its behavior."

Then Prof. Wilson after having stated the general functions of the centrosome, sums up in the following language: "These facts seem explicable only under the assumption that in these cases the centrosome or the substance which it carries gives an active stimulus to the cytoplasm, which incites the aster formation about itself and in words of Griffin disengages the forces at work in mitosis.

"The centrosome must, however, be something more than a mere division center, for on the one hand in leucocytes and pigment cells, the astral system formed about it is devoted, as there is good reason to believe, not to cell division but to movements of the cell body as a whole, and on the other hand, as we have seen, it is concerned in the formation of the flagella of the spermatozoon and probably also in the cilia of epithelial cells." You will see from this, that the centrosome is the party in the cell who seems to direct its several actions. It will also appear that the centrosome leads a separate life within the cell and multiplies by division in the same manner as the cell itself.

We shall now consider the functions of another important organ of the cell, the nucleus. This consists of a crowd or colony of skilled workers, who perform, look after and direct the different kinds of work. You will see later on, that the nucleus has a number of crowds or colonies specialized in different kinds of work. These colonies are called chromatin granules or grains and when lined up for division are called chromosomes.

Here is a description of the nucleus by Prof. Wilson: "The nucleus usually lies in the center, but as the outer wall thickens the nucleus moves toward it and remains closely applied to it throughout its growth, after which the nucleus often moves into another part of the cell. That this is not due simply to a movement of the nucleus toward the air and light is beautifully shown in the inner walls of the cell. The same position of the nucleus and movement of the nucleus is shown in all cases toward the place or near the place where the work is to be done."

I wish to make it clear to the reader that in the construction of any product in the body like bile, milk or digestive fluids, it is not produced by the gland, liver, etc., but by the individual cells of which they are composed; and that these individual cells are again made up of still smaller individuals, who do the actual work. You see the cell itself is a very highly organized being, just as the human body. The cells that build our body were not always in the habit of building themselves into large co-operative colonies like those we see now as animals, plants and trees. We find them still with other habits and methods of life, living the single and separate lives in ocean or fresh water and nearly all of them have some peculiar ways and actions that show wonderful skill and intelligence. The following is a description of the protozoa by Ernest Haeckel:

"Many of the aquatic protozoa have the power of autonomous and independent locomotion and this often has the appearance of being voluntary. Among the simplest fresh water protozoa are the (arcellina) little rhizopods that are distinguished from the naked amoeba by the possession of a firm envelope. They usually creep about in the slime at the bottom but in certain circumstances rise to the surface of the water. As Wilhelm Englemann has shown, they accomplish this hydrostatic movement by means of a small vesicle of carbonic acid, which expands their unicellular body like an air balloon; the specific weight of the cell body which is of itself heavier than water is sufficiently lowered by this. The same method is followed by the pretty radiolaria which live floating (as plankton) at various depths of the sea. Their unicellular body is divided by a membrane into a firm inner central capsule and a soft outer gelatine covering. The latter known as the calymma is traversed by a number of water-vesicles or vacuoles. As a result of an osmotic process carbonic acid may be secreted or pure water (without the salt of the sea water) be imbibed in these vacuoles; by this means the specific gravity of the cell is lessened and it rises to the surface. When it desires to make itself heavier and sink, the vacuoles discharge their lighter contents. These hydrostatic movements of the radiolaria attain by simple means the same end that is accomplished in the siphonophora and fishes by air filled and voluntarily contractile swimming bladders."

You will notice from this description by Mr. Haeckel, that this particular kind of cell understands all the laws of hydrostatics. He understands how to make and inflate his coat with a gas which lifts him to the surface of the water whenever he wants to go there for food or air. Whenever he gets through with his work at the surface of the water he allows the gas to escape and goes back to the bottom. I wish to call the reader's attention to these very complicated acts, showing that the cells which still live and lead a separate life possess a mind and intellect of a high order. You will notice that in the past some writers and scientists have made the remark that the cell appears to have a mind and a free will. Whether they have or not will be a question for the reader to determine for himself after having read this book. It has taken man a long time with a mind and free will to rise from the savage state to civilized life. Intelligence and free will have placed man where he is today and I believe that the same forces have been back of the progress and development of cell life which is back of man.

One writer makes the following statement about the single cells living in the sea: "A great number of classifications for the methodical distribution of these beings has been proposed but not one of them is altogether satisfactory. Some inhabit fresh and salt waters, serving as food for a great many other organisms or contributing by means of their calcareous or silicious skeletons to the formation of continents. Others live as parasites in the organs of animals and plants and induce more or less serious disorders in the constitutions of the organisms they have penetrated. Others again, acting like ferments produce important chemical modifications in organic matter in the course of decomposition."

You will notice from this statement that the cell is able to effect chemical modifications and changes in organic matter. This would seem to clearly indicate that the cell is familiar with the laws of chemistry. Mr. Ernest Haeckel makes the following statement about the cell: "When the cell-theory developed in the course of the last half century, the common anatomic ground work of all living forms was recognized in the cell, and the conception of the cell as the elementary organism led to the further belief that our own frame like that of all the higher animals and plants is a cell state composed of millions of microscopic citizens, the individual cells, which work more or less independently therein and co-operate for the common purposes of the entire community. This fundamental principle of the modern cell-theory was applied with great success by Rudolph Virchow to the diseased organism and led to most important reforms in medicine. The cells are in his view independent 'life units'." The reader will notice from this that the cell is now looked upon as a separate living individual and that in plants and animals cells live together in a co-operative community.

It seems to be the aim in nature to improve and develop. Improvement in a social way can come only at the cost of some of the liberties of the individual citizen. In proportion as society organizes itself and rises in scale, so does a shrinkage enter the private life of each of its members. In order to obtain the comforts and security of society the individual must abandon some of his personal liberties and work towards one purpose, the economic and political perfection of the race. Is it not wonderful that plants and animals happen to be perfect political and economic organizations of individual cells, just like our own organized society? Think of the countless billions of cells that have gone down in a terrific struggle with the elements in the past ages, in order to place civilized man in his secure and comfortable home.

We look down at the actions of the cell like a man looking at us from the sun or another world, never thinking that we are looking at our maker through a telescope made by him, the eye. It is no wonder that man, who was built and guided by the intelligence of the cell, was able to organize himself into a higher civilized life, because his builder had gone through all those experiences before. The customs, laws, virtues political and economic, now practised by civilized man have been practised by the cell for the past million years.

Before I go any further I shall quote some more passages by Prof. Wilson as to the general appearance, inner actions, and organs of the cell. The word protoplasm used now and then in reference to the cell is an old word which now is practically meaningless. However, as nearly as I can see, it means the whole cell. The following is by Prof. Wilson: "A minute analysis of the various parts of the cell leads to the conclusion that all cell organs, whether temporary or permanent are local differentiations of a common structural basis. Temporary organs such as cilia or pseudopodia are formed out of this basis, persist for a time and finally merge their identity in the common basis again. . . . . . The facts point toward the conclusion that the power of division not only of the cell organs but also of the cell as a whole may have its root in a like power on the part of more elementary masses or units of which the structural basis is itself built. . . . . . If such bodies exist they must, however, in their primary form, lie beyond the present limits of the microscope. . . . . The phenomena of cell division show, however, that the dividing substance tends to differentiate itself into several orders of visible morphological aggregates. . . . . . At the bottom of the series there must be masses that cannot be further split up without loss of their characteristic properties and which form the elementary morphological units of the nucleus. . . . . . These facts point unmistakably to the conclusion that these granules are perhaps to be regarded as independent morphological elements of a lower grade than the chromosome. . . . . . The smallest chromatin grains may successively group themselves in larger and larger combinations of which the final term is chromosome. . . . . . Whether these combinations are to be regarded as individuals is a question which can only lead to a barren play of words. Are these the ultimate dividing units as Braur suggests? When all these facts are placed in connection we find it difficult to escape the conclusion that no definite line can be drawn between the cytoplasmic granules at one extreme and the chromatin granules at the other. And in as much as the latter are certainly capable of growth and division, we cannot deny the possibilities that the former may themselves have or arise from elements having like power.

"A fragment of a cell deprived of its nucleus may live for a considerable time and manifest the power of coordinated movement without perceptible impairment. Such a mass of protoplasm is, however, devoid of the powers of assimilation, growth and repair, and sooner or later dies. . . . . . For these and many other reasons to be discussed hereafter the nucleus is generally regarded as the controlling center of cell-activity and hence a primary factor in growth development and the transmission of specific qualities from cell to cell and so from one generation to another."

In its simplest form the centrosome is a single minute granule, which may, however, become double or triple as in the white blood cells of the body, connective tissue cells or epithelial cells or skin cells. . . . . During the formation of the spermatozoon, the centrosome undergoes some remarkable changes and is closely involved in the formation of the contractile structures of the tail. The existence of cell organs, which have the power of independent assimilation and growth and division is a fact of great theoretical interest in its bearing on the general problem of cell organization for it is one of the main reasons that have led Boveries, Weisner and many others to regard the entire cell as made up of elementary self propagating units. Besides the active substance, the cell contains various lifeless bodies suspended in the meshes of the network. Examples of these are food granules, pigment bodies, drops of oil or water and excretory matters. . . . . . . Among the lifeless products of the body must be reckoned also the cell wall or membrane by which the cell bodies may be surrounded."

We shall now begin to consider the subject of reproduction. The cell multiplies by dividing itself into two parts. In this performance the directing center called the centrosome takes charge of the performance and looks after the details to see that the different specialized and skilled workers in the cell body are exactly equally divided. As this colony of skilled workers called the nucleus possesses the knowledge and experience handed down through past ages, it is very important that division be equal, so that each and all of the cells shall be possessed of the same amount of power, knowledge, skill and information. For this reason the centrosome or directing center of the cell takes full charge of this most important act.

The following illustrations, Figs. 4-5-6, by Prof. Wilson, will show how the performance of cell division looks through the microscope. The first act is division by the centrosome; next, the skilled workers, scattered pro-

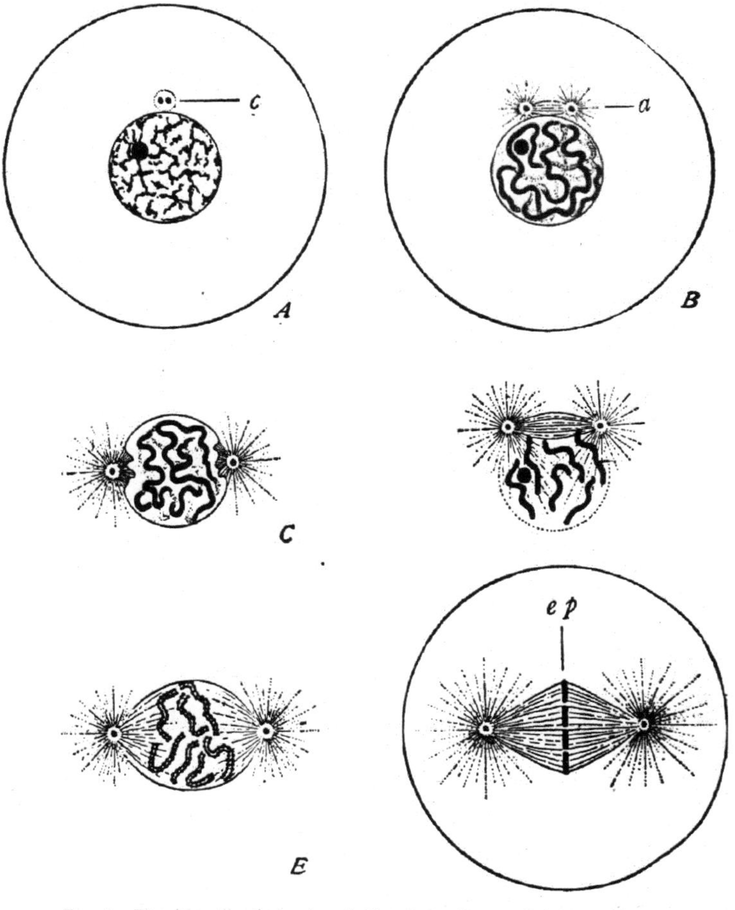

FIG. 4.—Diagrams of cell showing the first stages in act of division.—WILSON. A. Resting stage of cell. B. Beginning of division. C. Centrosome divided. E. Nucleus or crowd of skilled workers beginning to divide. F. Skilled workers lined up for division.

miscuously about in the nucleus, line themselves up in rows for division. Then the two centrosomes move over to opposite sides of the cell body. Then each centro-

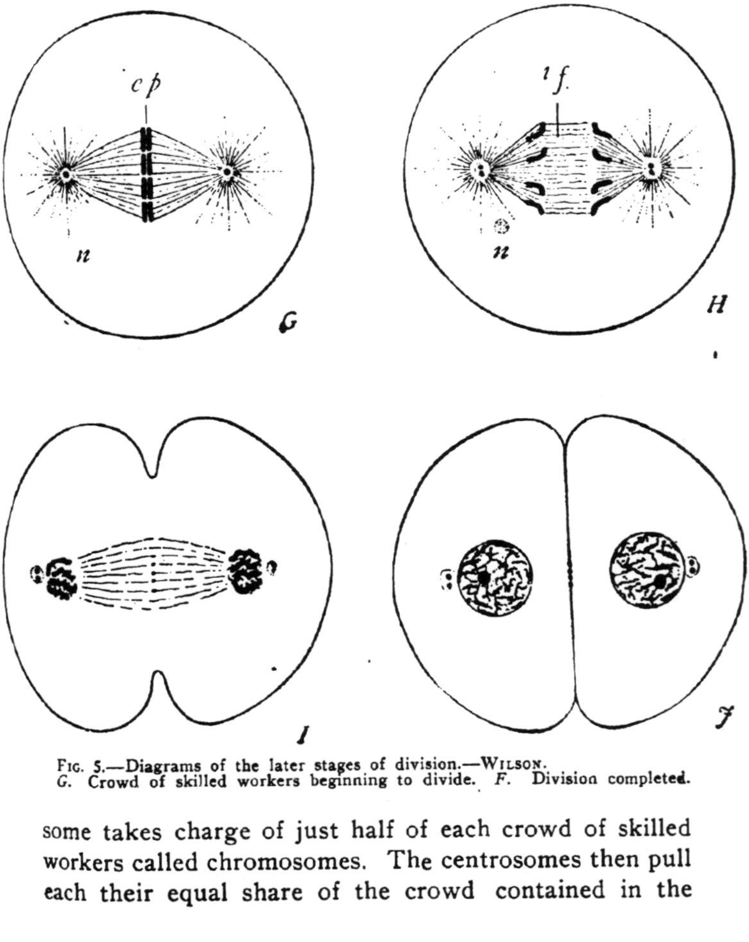

FIG. 5.—Diagrams of the later stages of division.—WILSON.
G. Crowd of skilled workers beginning to divide. *F.* Division completed.

some takes charge of just half of each crowd of skilled workers called chromosomes. The centrosomes then pull each their equal share of the crowd contained in the nucleus towards their side of the cell bodies, and finally the body itself divides, and each centrosome in that way obtains by this act just half of the original cell. Who is this most wonderful being, the centrosome, who directs these actions? Seen through the most powerful micro-

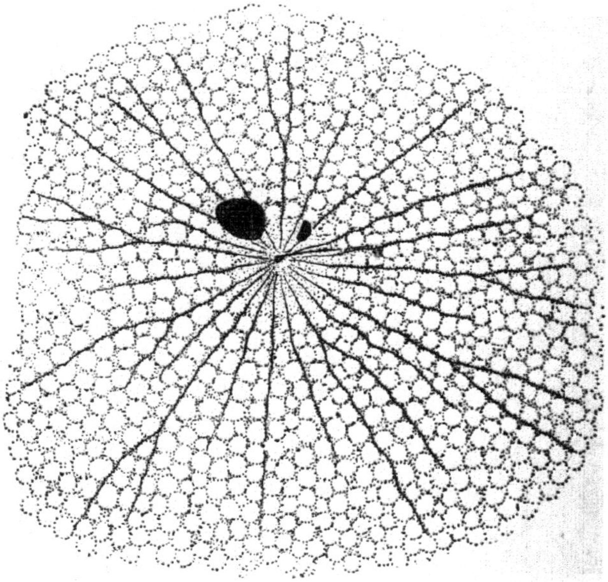

FIG. 6.—Centrosome highly magnified.—WILSON.

scope, he looks like an individual who is in touch with all the other individuals around him by some method or means that is not yet understood. He seems to be connected with everything in every direction as you will see from the illustration.

In reference to the work of the centrosome Mr. Wilson makes the following remark: "The division of the cytoplastic granules must remain a quite open question, yet we should remember that in dividing plastids of plants cells are often very minute, and that in the centrosome we have a body no larger in many cases than a microsome, which is positively known to be in some cases a persistent morphological element, having the power of growth, division and persistence in the daughter cells. When we consider the analogy between the centrosome and the chromatin grains, when we recall the evidence that the latter graduate into the oxychromatin granules, and these in turn into cytomicrosomes, we must admit that Burke's cautious suggestion that the whole cell might be a congerie of self propagating units of a lower order is sufficiently supported by facts, which constitute a legitimate working hypothesis."

You will see from this that the general opinion of scientists is that the cell is a colony of still smaller cells or beings; and that the centrosome is simply one of these smaller cells specifically in charge and organized to be the general director and manager of the whole organization we call the cell. Figure 7 of the centrosome shows clearly that it is in contact with every part of the body of the cell. The reader can clearly see that the individuals organized together to make the complete whole we call cell are too small to be clearly seen and to have their individual actions studied.

Now we come to consider the most wonderful performance in plant and animal life, which are the reproductive actions of the cells that build plants and animals. As the reader probably knows, all life we see, such as plants and animals, begins as one single cell. In reference to this point Prof. Drummond makes the following remark: "The embryo of future man begins life like the primitive savage in a one room hut, a single simple cell. The cell is almost microscopic and round in shape. An outer covering almost as transparent as glass surrounds this little body and in the interior imbedded in the protoplasm lies a bright globular spot; in form, size and composition there is no difference in this cell and that of any other animal. The dog, elephant, lion and a thousand others begin their lives the same way. At an earlier stage before it has taken on its pellucid covering, this cell has affinities still more astounding for it is a fact in modern science that the first embryonic abodes of moss, fern, pine, shark, crab, lizard, lion and man are so exactly similar that the highest power of microscope and mind fail to trace the smallest distinction between them."

You will see from this statement by Mr. Drummond that what I have told you heretofore is true, that everything starts from one single cell. That they are all alike except as to outside covering and size. The cells that build man, mouse or plants look alike in the same way that a shoemaker and a preacher look alike, but their knowledge and experience in life have been different. For that reason the work they do and the structures they build will be different, depending on what their education and experience have been. The actions of a living being are based on his knowledge, derived from previous experience.

To return to the question,—where does this first cell come from: When the cells build the animals, special organs are built in the male and female where the cell is prepared, educated and dressed up; these are the sex organs. In a certain place in the female bodies, arrangements have also been made to furnish these cells with food and building material, so they can multiply and build another body. This place is called the womb. Now how do they prepare for this act of meeting in the womb for the purpose of building a new body? In the overies of the female the cell prepares for the occasion in this way: It discharges or dispenses with half of the skilled workers located in the nucleus, and also its centrosome or general manager and provides itself with food

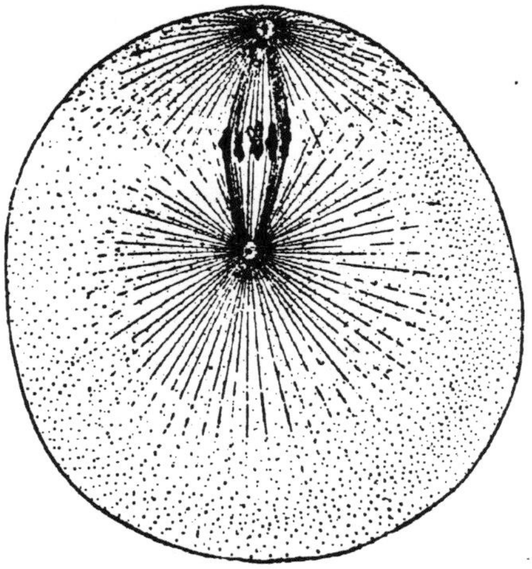

FIG. 7.—Structure of the centrosome in the polar asters.—WILSON.

enough to last several days. With this food she goes into the womb and waits for the cell to come from the male. How is he prepared? He also discharges or abandons half of his skilled workers in the nucleus but keeps the general manager or centrosome, so when the two meet and join their forces together, there will be a complete cell with a full nucleus, centrosome and also food enough to last them until connections have been built up in the womb with the body, where they can obtain and be furnished with such food and building material as will be required to build the new structure. I must further describe the male cell, who makes elaborate preparation to be able, not only to meet the female cell, but to be able to be the first one to reach her in the mighty race and struggle that takes place between a thousand or more of them in a dash for the goal.

The male cells, or spermatozoons, who prepare to meet the female cells in the womb, change their form so completely that no one not familiar with them could possibly recognize them as a human cell. They were thought to be parasites when first discovered and not in any way connected with or a part of the body. The male cell is reconstructed to effect a certain purpose, which is to push its way through obstacles in a competitive race that is to take place by thousands of individuals at the same time. To be swift and effective, the centrosome is placed in front, protected by a hard point; next, back of him is the entire crowd of skilled workers known as the nucleus; then back of them there is a powerful tail or propeller, constructed from the common workers and laborers of his whole body, which is designed to push the male cell with the centrosome in front, directing the course to the place where the female cell is waiting for him in the womb.

The following is a description from Prof. Wilson of the male cell generally called the spermatozoon: "'In its more usual form the animal spermatozoon resembles a minute, elongated tadpole, which swims very actively about by the vibrations of a long slender tail. Such a spermatozoon consists typically of four parts. Fig. (8).

First—The nucleus, which forms the main portion of the head and consists of a very dense and usually homogeneous mass of chromatin staining with great intensity. It is surrounded by a very thin cytoplasmic envelope.

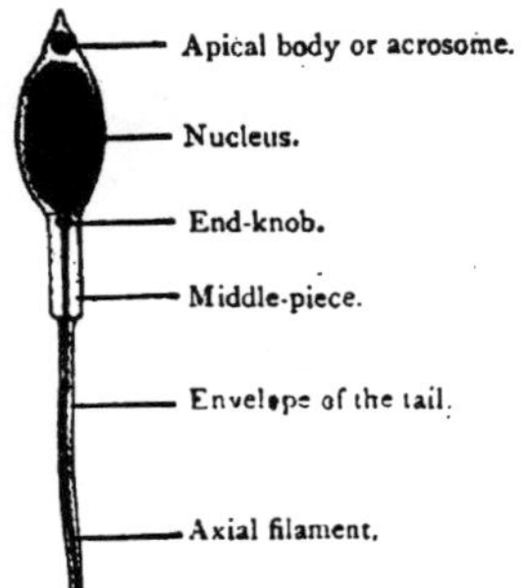

Fig.

FIG. 8.—Diagram of the flagellate spermatozoön.—WILSON.

Second—An apical body, or acrosome lying at the front end of the head, sometimes very minute, sometimes almost as large as the nucleus and in some cases terminating into a sharp spur by means of which the spermatozoon bores its way into the ovum.

Third—The middle piece, or connecting piece, a large cytoplasmic body lying behind the head and giving attachment to the tail from which it is not always distinctly marked off. At its front end it is in some forms (mammals) separated from the nucleus by a short clear region, the neck.

Fourth—The tail, or flagellum, in part at least, a cytoplasmic product, developed in connection with the centrosome. From a physiological point of view we may arrange the parts of the spermatozoon under two categories as follows:

First—The essential structures, which play a direct part in fertilization; these are: (a) The nucleus which contains the chromatin. (b) The middle piece, which either contains a formed centrosome or a pair of centrosomes (end-knob), or is itself a metamorphosed centrosome. This is probably to be regarded as the fertilizing element par excellence, since there is reason to believe that when introduced into the egg, it gives the stimulus to division.

Second—The accessory structures, which play no direct part in fertilization, viz; (a) The apex, or spur, by which the spermatozoon attaches itself to the egg or bores its way into it. (b) The tail, a locomotor organ, which carries the nucleus and centrosome, and as it were, deposits them in the egg at the time of fertilization. There can be little doubt that the substance of the flagellum is contractile and that its movements are of the same nature as those of ordinary cilia. Ballowitz's discovery of its fibrillated structure is therefore of great interest as indicating its structural as well as physiological similarity to a muscle fibre.

"Tailed spermatozoa conforming more or less nearly to the type just described are with few exceptions found throughout the metazoa from the coelenterates up to man."

The reader will see from this statement, Fig. (8), by Prof. Wilson, that it appears clearly the male cell is a structure built up and composed of a multitude of still smaller primordial beings; that the central part of the cell, the nucleus, contains those who are skilled and understand the work that has been done and is to be done in the future, and that the centrosome is the directing center and general manager whose duty it is to guide the actions of all; that the main body of the cell which is called cytoplasm, contains the common laborers and servants, who work and act from instructions and orders given by the skilled workers or centrosome. We see that the male cell in this race to find the female cell employs the entire force of common laborers in the body of the cell to push and carry him to the desired place. Consider for a moment this preparation of thousands who are willing to prepare for this race, where all but one must fail to reach the goal, and all but one must perish. However, it is a fair race, based on the law of evolution, that the best man in a struggle for existence will win. In this act, as in all other schemes in nature, where there is intelligence there is authority, design, purpose, wisdom and victory. Here it seems that microscopic and mysterious beings we call the centrosomes are able to direct all the other microscopic beings in the cell to subjugate and organize matter, so as to create for themselves an existence on the planet. It appears that the centrosome is the real source of will and intelligence in the cell, just as the brain cells are the source of intelligence in animals and man.

This male cell of animals is so very similar in shape and actions to male cells or spermatozoons of plants and insects, and also cells living a single and separate life in the water, that I shall quote a description of them from Ernest Haeckel.

"The same may be said of the traveling spores of many of the algae and of the most remarkable of all ciliated cells, the spermia or spermatozoa of plants and animals. As a rule they are cone shape, having an oval or pear shape, (though often also rod shape) head, which tapers into a long and thin thread. When their lively movements were first noticed in the male seminal fluid (each drop of which contained millions of them) two hundred years ago, they were thought to be real independent animalcules like the infusoria and so obtained their name of seed animals (spermatozoa). It was a long time (60 years ago) before we learned that they are detached glanular cells, which have the function of fertilizing the ovum. It was discovered at the same time that similar vibratory cells are found in many of the plants (algae mosses and ferns). Many of the latter (for instance, the spermatozoids of the Cycadea) have instead of a few long whips, a number of short lashes (cilia) and resemble the more highly developed ciliated infusoria.

"The ciliary movement of the infusoria is held to be a more perfect form of vibratory movement, because the many short lashes found on them are used for different purposes and have accordingly assumed different forms in the division of labor. Some of the cilia are used for running or swimming, others for grasping or touching and so on. In social combination we have the ciliated cells of the ciliated epithelium of the higher animals—for instance, in the lungs, nostrils, and oviducts of vertabrates."

The reader will clearly see from this description by Mr. Haeckel, that the male cell of man, animals and plants is in no way different from the other single cells leading a single life in the water. That he is a very active animal with organs of touch, locomotion, and a great number of others to assist him in his life. Many male cells of plants and animals swim about in the water for a long time and long distances in search of the female cell. It shows clearly that at the start we are just what Prof. Drummond stated, "Apparently no different from plants insects, and animals." These tiny microscopic beings will of their own free will go ahead and multiply into groups, and tie and arrange themselves into the different shapes and forms we afterwards call plants, animals or man.

By this time, I think the reader must see that the cell is the one who is at the base and is the cause of life that we see, as in plants and animals. He is the cause of all living structures. He multiplies by division, as heretofore described, and with the building material at hand or furnished by other cells, the work continues until the plant or animal is completed. We see him move about from place to place in search of food and material for his structure just as other animals do.

We know that he requires food, air and water just as we do. We know that he understands how to select the right kind of material at the right time and with it, build the most intricate and complicated structures. We know this because we see him do it and from these facts, we must naturally conclude that he acts in a similar manner to other animals including man. We find that he acts for a purpose. When man acts for a purpose, we say he is intelligent. When a person is able to produce any structure by reason of being intelligent, we mean that he is instructed or skillful in the particular work which he does. When a large building is being constructed, it seems in the distance to simply grow larger from day to day; it appears in the distance to take form slowly and imperceptibly. If we could never get near enough to the building to clearly see what was going on, we would be compelled to say that it simply grew. Why? Because we had not eyes powerful enough to see what was being done; but with the aid of the telescope we could increase the power of our eyes to see more clearly and determine the facts, that the building was not being formed by magic but by the work and industry of a number of human beings. So it is in every day life, we see a tree or plant or animal grow. That is as far as we can see. By increasing the power of our eye with a miscoscope, we can see that it does not simply grow, as by magic, but that it is being constructed in the same manner as a building would be by man. Every part of the plant is occupied by active individuals, continuously at work, feeding themselves, multiplying and building the plant or animal or its parts.

We shall again consider what takes place in the process of division of the cell. All modern authority now seems to agree that the cell itself is a colony of beings, organized to work together for the mutual benefit of all. From the fact that not only the cell itself, but also each specifically skilled crowd of workers contained in the nucleus divides exactly in two equal parts, so that each cell gets exactly its equal half of the skilled workers, makes it clear that those individuals of the cell who are concerned with the work of building and keeping a record of past experiences, which we call memory, are at all times kept double or in duplicate. After the crowds have divided as takes place in the division of the cell, another division will not take place until that crowd has again doubled. It seems clear to me that those chromatin granules, as they are called, are individuals or primordial beings possessed of the special knowledge and skill of building those structures from whence they came. The fact that the centrosome, who takes charge of the act of division, causes such division to be exactly equal, tends to show a definite purpose to divide the specifically skilled working force equally. How this force of individuals is again able to grow and increase in the cell to its full number and size, the microscope is not yet powerful enough to disclose to us. However, we can clearly see that it does again increase and grow to its original number and size.

In the act of fertilization where the male cell prepares to meet the female cell, it is clear that each one knows what the other is going to do and acts accordingly. The female cell must know that the male cell will bring a centrosome or general manager and the male cell must know where to look for the female and also have some idea what she will have and what she looks like. They must have the faculty of memory, skill and intelligence to do these things. Every act connected with this performance of the male and female cells of preparing for and meeting in the womb, is an intelligent act requiring judgment and will, which must be based on memory. We do not know how the cells in our brain can remember past experiences, but we know it to be a fact. We know every cell in our body including our muscles must also be possessed of this faculty of memory, just as the cells composing our nerves and brain are, if not we could learn nothing from practice nor form a habit. Habit is based on the memory of the cells of that part of the body trained or practising. By repetition the acts are fixed in the memory of the cells in the muscles that take part in the act, and finally they are remembered and then we call them habit.

We have seen that the cell is an individual again composed of individuals called granules; that these granules, usually called chromatin granules, group themselves into crowds who have their own specific work to do in the same way that in our body we have special colonies or crowds of cells, who have special work to do and are called organs, like the heart, liver, lungs, stomach, etc. So in the cell the skilled crowds, who each have their special work to do in the body of the cell, are called organs. There is the centrosome, who is the directing center or general manager of the whole cell body. Then there is the nucleus, that contains an enormous number of crowds, each skilled in his particular kind of work, as we shall see later. Then there is the group that has nothing to do but make starch for food. This crowd of special workers are found only in the plant building cells and those plants who are able and understand how to make their own food out of the raw material of earth, air and water. Then there is the crowd of general workers, which you might call common laborers, who compose the main body of the cell. How do the cells communicate with each other? That they have some method of certain and rapid communication is clear, and is evidenced by the rapidity with which sensations and ideas are transferred from one place of the body to the other. It has been demonstrated that the nucleus in one cell will take charge of the common laborers and workers in another cell, who have been deprived of their nucleus. Here is Prof. Wilson's statement of those experiments: "If correct, these experiments give clear evidence of transference of physiological influences from cell to cell by means of the protoplasmic bridges, showing that the nucleus in one cell may thus control the membrane-forming activity in an unnucleated fragment of another cell." This is very significant in showing with what loyalty they co-operate and work together for the general welfare of the entire colony or body.

I must here explain to the reader more about the reproductive cells of plants and animals,—how they differ from those cells that build the body. They are called germ cells and as you remember, the male cell is called the spermatozoon and the one that comes from the female is called the egg. However, the reader must not forget that the egg and spermatozoon are cells, in every way just alike, except that they are built and prepared to perform different kinds of work and for that reason, they look different. The egg or female cell is prepared to just sit and wait for this spermatozoon with food for both to last them until they can be connected with the nourishment from the body. The spermatozoon is built with a powerful propeller with which he is enabled to push his way through obstacles and obstructions in a mad competitive race with thousands of others, who all start at the same time in search for the same object, the female cell. This act of the spermatozoon and female cell meeting and joining is called fertilization. You remember that the nucleus, which always contains the same number of specifically skilled crowds of workers, always divides exactly in two whenever the cell reproduces itself or multiplies by division. Now it is also a singular fact that the cell from the female and also from the male discharge just half of the specifically skilled crowd of workers in the nucleus, which are called chromosomes, so that when the male and female cells get together and find each other, and the two join together to form one cell, the two together will be a complete cell, and will have a full number of specifically skilled workers, and the correct number of chromosomes. The body cells or cells that build the body are called somatic cells, and are always a little different from the germ cells. Whenever these cells, which build the body, begin their task,—as a liver cell, or muscle cell,—they first discharge or dispense with a part of their nucleus or specifically skilled workers, according to the kind of work they have to do. These facts are very significant in that they prove that the nucleus contains groups of workers specifically skilled in all the departments of work pertaining to the building of the body. When a cell takes charge of the work of building and maintaining the liver or lungs it retains only that group of specifically skilled workers who understand that particular line of work, and discharges all the others. In reference to these facts, I will here quote some more from Prof. Wilson:

"The germ from which every living form arises is a single cell derived by division of a parent cell of the preceding generation. In the higher types the germ cells are more or less definitely organized in groups supported and nourished by somatic cells, specially set apart for that purpose, and forming distinct sexual organs, the ovaries and spermaries, or their equivalents. Within these organs the germ cells are carried protected and nourished and here they undergo various differentiation to prepare them for their future functions. The structural difference thus brought about between the germ cells is, however, only the result of physiological division of labor. The female germ cell or ovum supplies most of the material for the body of the embryo and stores the food by which it is nourished. It is therefore very large and contains a large amount of cytoplasm laden with food matter, and in many cases becomes surrounded by membrane or other envelopes for the protection of the developing embryo. On the other hand the male germ cell or spermatozoon contributes to the mass of the embryo only a very small amount of substance comprising as a rule only a single nucleus and a very small quantity of cytoplasm. It is thus relieved of the drudgery of making and storing food and providing protection for the embryo and is provided with only sufficient cytoplasm to form a locomotor apparatus by which it seeks the ovum. It is therefore very small and performs active movements. The plant ovum which is usually known as the oosphere shows the same general features as that of animals. The flagellum or tail is merely a locomotor organ, which plays no part in fertilization. Its most characteristic feature is the axial filament, which is composed of a large number of parallel fibrillae like a muscle fibre. Both the ova and spermatozoon take their origin from cells known as primordial germ cells, which become clearly distinguishable from the somatic cells at early period of development and are at first exactly alike in the two cases. Moreover, from the outset the progenitors of the germ cells differs from the somatic cells not only in the greater size and richness of chromatin of its nuclei but also in its mode of mitosis, for in all those blastemers destined to produce somatic cells, a portion of the chromatin is cast out into the cytoplasm where it degenerates and only in germ cells is the sum total of the chromatin retained. Only the germ cells receive the sum total of the egg chromatin handed down from the parent. All of the somatic cells contain only a portion of the original germ substance. The original nuclear constitution of the fertilized egg is transmitted as if by law of primogeniture, only to the one daughter cell and by this again to one and so on. While in the other cells, the chromatin in part degenerates, in part is transformed so that all of the descendants of these side branches receive small reduced nuclei. The number of chromosomes arising from the germ nuclei is always the same in both and is one-half the number characteristic of the tissue cells of the species. The two nuclei do not fuse, but only place themselves side by side and in this position give rise each to its own group of chromosomes."

"The difference between the two sexes is probably due to the physiological division of labor between the two germ cells, the spermatozoon being motile and very small, while the egg contains a large amount of protoplasm and yolk. The evidence is steadily accumulating, that reduction is accomplished by two maturation divisions throughout the animal kingdom, even in the unicellular forms. The one fact of maturation that stands out with perfect clearness and certainty amid all the controversy surrounding it, is a reduction of the number of chromosomes in the ultimate germ cells to one-half the number characteristic of the somatic cells. It is equally clear that this reduction is a preparation of the germ cells for their subsequent union and a means by which the number of chromosomes is held constantly in the species."

You will notice that the actions of the individuals which compose the cell show that they are very careful not to lose any of the crowds of specifically skilled workers; that great care is exercised to see that the division is exactly equal. The one who always takes charge of the act of division is known as the centrosome. You will also notice from Mr. Wilson's statement, that the group of workers in the nucleus do not "fuse," as we used to think when we did not have microscopes powerful enough to see what became of them. You will notice that they simply remain side by side and apparently do so during the entire life of the individual. From the evidence before us now it begins to look quite clear how the cell can and does build a structure like a plant or animal. A cell is a combination of a multitude of highly organized and specialized primordial beings. The cell contains in itself numerous crowds of specifically skilled workers, numbering thousands or millions. How these primordial beings that make up the cell actually multiply and grow, we have no means of knowing nor microscope to tell us. You see life is based on organization and specialization. When the cell multiplies by division there is simply an equal division of the colony of specifically skilled workers in the cell, which is called the nucleus. In this way, you can clearly see, that the knowledge, skill and experience possessed by the cell, or more correctly speaking, by the individuals composing the cell, and which they have accumulated through the past ages of experience, is handed on to posterity and preserved.

It is a singular thing how everything works out according to a preconceived plan; how exactly one part is made to meet or fit into another part in the course of development. The spermatozoon knows what the female cell will have to start in life with so he comes to meet her prepared accordingly. The female cell seems to know what the spermatozoon will bring, so she leaves those same things behind. She knows what he will have to contend with in his struggle to reach her and that he cannot be bothered with carrying a food supply, so she provides the food. She seems to know how much she will need and how long it will take the body cells to connect them with food supply from the body. In several species special food carriers are provided, which are called nurse cells. They stay with the germ cells and furnish them with food until they get started in life. Here is what Prof. Wilson has to say about them:

"As the primordial germ cells enlarge and form the mother cells of the egg, they almost invariably become intimately associated with neighboring cells, which not only support and protect them but also serve as a means for the elaboration of food for the growing egg cells. In most cases, as ovarian development proceeds, definite association is established between the egg and surrounding cells. As a rule the material elaborated by the nutritive cells is passed into the egg either in solution or in the form of granular protoplasmic substance. In some form each egg is accompanied by a single nurse cell attached to its side with which it floats free in the body cavity. The nurse cell is at first much larger than the egg cell. The egg cell rapidly grows, apparently at the expense of the nurse cell, which becomes reduced to a mere rudiment, attached to one side of the egg and finally disappears. In mysostoma, the young egg is accompanied by two nurse cells, one at either end."

This shows clearly that the germ cells, as they start out from one body with the intention of building another and new body, are generally provided with food to give them a start in life; and in many cases separate cells or servants are provided who go with them and carry the food for them. The reader must be able to see at this time that the cells in all the details of their inner life act precisely as we do. They act with a purpose and according to a preconceived plan. Why should they act differently? The ideas that direct our actions are nothing more than the ideas of cells located in our head, who have charge of that particular work in our bodies of thinking and directing our actions.

There must have been a time in the past million years when those primordial cells that make up the cell and which they call chromatin granules, lived separate and single lives alone in the world, just as a great number of species of the single cell are doing today. You might say that from the way it looks, the cell is not the real builder but that the real builder is the primordial cells that have associated together and organized this individual we call cell. However that may be, the cell is clearly a separate living individual or animal regardless from what he is organized or of what composed. Ernest Haeckel states, "The cells are individual life centers, and the unified life of the whole man is the combined result of the work of his component cells. In this way the cells are the real life units of the organism. Their individual independence is at once seen in the permanently unicellular protists of which several thousand species are already known to us."

It appears that the cell has organs with which it is able to see, feel and hear. Still I do not see how they can have these organs in the same sense that we consider hearing and seeing. Here is what one scientist has to say about the cell possessing organs of sight and hearing: "It is impossible to believe that these organs are not eyes for they have the same structure as eyes of comparatively higher classes of animals, such as certain worms, tubularia, rotifers, lower class crustaceans, etc. All these organs are similarly formed of a small crystalline globule inclosed in a small mass of pigmentary matter. The identity of structure naturally leads to the assumption of the identity of functions."

Ernest Haeckel claims for the cell a definite mouth and annus. Here is what he has to say:

"The great class of rhisopods is distinguished by the fact that their naked plasma bodies can take in ready formed solid food at any point of the body. On the other hand most of the infusoria have a definite mouth opening in the outer walls of their unicellular bodies and sometimes a gullet tube as well. Besides this cell house we usually find also a second opening for the discharge of indigestible matter, a cell annus."

Prof. Haeckel has also been able to see the movements

of the primordial cells inside of the cell and this is what he says in reference to their appearance:

"In the rhizopod, the remarkable protozoan who as a unicellular organism sheds so much light on the obscure wonders of life, we notice a curious streaming of the granules in the living plasm. Within the cytoplasm of the amoeba particles travel up and down in all directions."

The following is Prof. Haeckel's observation in reference to how they act in conjugation:

"When two ciliated Infusoria conjugate, they place themselves side by side and connect for a time by means of a bridge of plasm. A part of the nucleus of each has already divided into two portions, one of which functions as the female standing nucleus and the other as the male traveling nucleus. The two mobile nuclei enter the plasm bridge and move through it, pushing against each other into the body of the opposite cell; they then coalesce with the deeper lying standing nucleus. When the fresh nucleus has thus been formed (by amphimixis) in each of the copulating cells, they again separate. The two rejuvenated cells have once more acquired the power to propagate for a time by division."

Mr. Binet, who believes the cells show a psychic phenomena, makes the following remarks: "The sexual elements and especially the spermatozoid of all unicellular organisms are certainly the ones which show the most highly developed psychical functions: the act of seeking and approaching the ovule, which is frequently situated at quite some distance from where the male element is deposited; the length of road to be traveled; the obstacles to be overcome; all point to faculties in the spermatozoid that are not explainable by simple irritability."

The reader by this time will see that not only has the cell all the special sense organs possessed by animals in general but in addition has others, the purpose of which is not yet understood; and that it performs all the actions and functions of life performed by animals. It must begin to appear to the reader at this time that the cell must have a mind and intelligence similar to our own. However, the scientists of today do not seem to so consider it. Therefore, I must use these statements of others in reference to the actions of the cell to demonstrate my contention.

We shall now consider their method of nutrition and respiration. The reader must remember that the cell is a perfect and complete animal, that it must have food, air and protection from the elements in the same manner as a human being. While it has been quite conclusively shown heretofore that the cell must have oxygen from the air and also food in order to exist, I shall quote a few statements from scientists to show the actions and requirements of the cells. "There seem to be three classes or methods by which cells obtain their food, viz:

First—The plant cells that make their own food from the raw material of earth, air and water through chemical action brought about by the aid of sunlight.

Second—Saprophytic beings who live on decayed or dead matter.

Third—Those that live on other lives. The second class are called scavengers and the third class parasites." I shall here quote Mr. Binet, who makes the statement, that the actions of the cell show choice and discretion; and this is what he has to say in reference to their mode of nutrition:

"Nutrition by endosmosis, or saprophytic. The organism nourishes itself by absorbing through the whole surface of its body, liquids containing the products of vegetable or animal decomposition. Saprophytic beings are found in putrid waters or in infusions.

"There is now a last mode of nutrition, of which we shall treat in minute detail; viz., animal nutrition, where the micro-organism seizes solid alimentary particles and nourishes itself after the fashion of an animal, whether it be by means of a permanent mouth or by means of an adventitious one improvised at the moment of need.

"Animal nutrition requires very remarkable psychological faculties in the organisms practising it. These manifestations of psychic life, the progressive complexity of which we intend to trace in starting from the simplest protozoic forms and arriving at the higher, prove that these animalcula are endowed with memory and volition.

"The micro-organisms do not nourish themselves indiscriminately, nor do they feed blindly upon every substance that chances in their way. Also when they ingest food through some point or other of their bodies, they understand perfectly how to make a choice of the particles they wish to absorb. This choice is sometimes quite well defined, for there are species which feed exclusively upon particular foods. Thus there are herbivorous Infusoria and carnivorous Infusoria. Among the herbivorous ones may be classed the chilodons, which feed upon small algae diatomaceae and Oscillaria. The paramecia live principally upon bacteria, the leucophrys is a specimen of the carnivorous class. It devours even the smaller animals of its own kind."

"The Bodo caudatus is a voracious flagellate possessed of extraordinary audacity. It combines in troups to attack animalcules 100 times as large as itself as the colpods for instance, which are veritable giants when placed along side of the Bodo. Like a horse attacked by a pack of wolves, the colpod is soon rendered powerless; twenty, thirty, forty bodos throw themselves upon him, eviscerate and devour him completely (Stein)."

Here is what Mr. Haeckel has to say about the parasitic habits of some cells.

"By parasites in the narrower sense, we understand in modern biology only those organisms which live on others and derive their nourishment from them. They are numerous in all the chief divisions of the plant and animal kingdoms and their modifications are of great interest in connection with evolution. No other circumstance has so profound an influence on the organisms as adaptation to a parasitic existence.

"I have already spoken of the many peculiarities of metabolism in the ubiquitous bacteria; while many of them cause putrefaction, they at the same time feed on the parts of other organisms which have died. The fungi feed for the most part on the decayed remains of plants and the product of putrefaction which accumulates on the ground. In this character of scavengers they play the same important part on land as the sponges do at the bottom of the sea."

I must here tell the reader something about the cell colony called the volvox:

These cells bunch themselves into colonies of about twelve thousand individuals. Some are male and female and some only workers or neutrals. The common laborers or neutral cells are all provided with hands which they use to push and propel themselves through the water backwards or forwards like a raft or submarine. Five thousand like the galley slaves of old paddle the water with their arms at the same time and in that way transport the other seven thousand males and females to such places as they may choose to go. Just consider this for a minute: a boat or raft pushed through the water by five thousand individuals, all working under the command of some one cell directing their course. They must necessarily work under the direction of one cell or else how could all these five thousand individuals push or pull in unison. How could they all pull or push together the same instant in order to go either forward or backward unless there was someone in command to give the orders when to go either forwards or backwards. This shows clearly how the cells are able to organize themselves so as to work together in harmony to affect their purpose. Still it is but a very crude organization as compared with the high state of organization and co-operation practised by the plant and animal building cells. Here is what Mr. Binet has to say about this colony of cells called the volvox:

"In the genus of volvox colonies are found of which the structure is very complicated. Such are the great green balls formed by the aggregation of diminutive organisms, which form the surface of the sphere and are joined together by their envelope. They have each two flagella which pass through the inclosing membrane and swing unimpeded on the outside; the envelopes each tightly holding the other form hexagonal figures exactly like the cells of a honeycomb. Each volvox is at liberty within its own envelope, but it projects protoplasmic extensions, which pass through its cuticle and place it in communication with its neighbor. It is probable that these protoplasmic filaments act like so many telegraphic threads to establish a net work of communication among all the individuals of the same colony. It is necessary, in fact, that these diminutive organisms be in communication with each other in order that their flagella may move in unison and that the entire colony may act as a unit and in obedience to a single impulse. The number of micro-organisms constituting a volvox colony is quite considerable; as many as twelve thousand have been counted.

"It was upon analogous phenomena that Gruber based the existence of a diffused nervous system in the stentors. The same line of reasoning may be followed in the case of the volvox. Since unanimity of movement is demonstrable among twelve thousand micro-organisms constituting a colony, it must be inferred that their movement is regulated by the action of a diffused nervous system present in the protoplasm. This conclusion is all the more interesting from the fact that these volvox are vegetable micro-organisms. In the dioecian volvox the female cellules and male cellules are joined together by themselves in separate colonies. When the time of fecundation arrives the male cellules scatter and proceed to conjugate with the female cellules. The colony which bears the female cellules also contains neutral cellules which are not designated for fecundation; the latter simply perform a locomotive function; equipped with one eye and two flagella, they are intended to move the great colonial ball; they are the oarsmen of the colony. The volvox male, female and neutral all seek the light, whether solar or artificial, and settle near the surface of the water. As soon as the female colonies have been fecundated, the oospores change their color, they turn from green to an orange yellow. At this time the colony is seen to draw away from the light and to disappear from the surface of the water. This change of position is affected by means of the vibratile cilia with which each neutral cell is furnished and which projects beyond the gelatinous sphere. Now as no change of color or form is noticed in the neutral cells after fecundation, it may be asked from what cause they flee from the light which they formerly sought."

You notice he ends by asking why they leave the light

which they formerly sought. Now to me that is very clear. These volvox cells are of the plant species. They understand how to make their own food and building material from the raw material at hand by the aid of sunlight. They row their raft into the sunlight now and then in order to make food consisting generally of starch and such other material as they need and when they get through, they move into the shade to rest up and enjoy themselves.

There is one method practised in reproduction by the smaller cells like bacteria and plant building cells, that is, the forming of spores, which I must mention here briefly. A spore is a cell protected with a hard covering against wind, weather and climate, wherein it can remain until favorable conditions shall arise to perform the ordinary functions of life or to begin the building of a new structure. Some spores, however, are able to swim about actively until they find their mates or places to locate and build their new structures, whether plant or animal.

The following is from our High School Botany in regard to these individuals called spores:

"Any one of these cells may produce within itself a single large swimming spore, which escapes from the mother cell into the water. At its more pointed clear end, there is a little crown of cilia by means of which it swims about rapidly. These spores finally anchor themselves and each one produces a new filament. It has been observed that these small swimming cells come together in pairs and fuse, each pair thus forming one new cell. The cell thus formed passes through a resting period (usually during winter) then begins to grow and finally produces four swimming spores, each of which is able to produce a new filament of ulothrix. Here is evidently a third method of reproduction, which is peculiar in the fact that two special cells unite to form the spore that produces the new cell.

"When these cells formed by internal division escape from the mother cell into the water, it is discovered that they are able to swim about by the lashing movements of four cilia, that appear in a cluster at the pointed end.

"A special cell thus set apart for reproduction is called a spore, and spores that swim are distinguished as swimming spores. A very important fact about ulothrix therefore is that it reproduces not only by vegetable multiplication but also by swimming spores.

"In other cells of the same filaments, or in cells of filaments under different conditions, the same formation of cells by internal division may be observed, but the contained cells are smaller and more numerous. When they escape it is discovered that they also are ciliated swimming cells but since they do not produce new filaments, it is evident that they are not swimming spores."

I think I have now covered in a general way the most important features connected with the inner life of the cell. I might mention that the ability of the cell to provide itself and build around itself protective coverings and armor of different materials and designs, from a soft flexible covering to one as hard as flint, has been the stepping stone by which the cell has been able to build the many peculiar structures, and to display the marvelous architecture which we shall find when we come to consider the living structures, in a chapter devoted entirely to that subject.

CHAPTER 4.

THE LIVING STRUCTURES.

The crust of the earth discloses to us like the leaves in a book an enormous number of animals and plants that have lived in the past but are now extinct and no more. Every layer and stratum is a page in the history of the life that has existed on this planet in the past. Just what was the cause of failure and its extermination in some cases is not easy to tell. However in the evolution of living structures, it is the same as in the evolution of other structures and those produced by man, the best will be retained and the inferior rejected.

The discovery by some animals of how to make feathers with which to cover their bodies and wings with which to fly, was a great invention and improvement, which caused a great change in the flying animals. The living structures that have existed in the past show the same gradual evolution and improvement by new inventions and discoveries as the evolution of man from his savage state up to the present time.

Mr. Haeckel states, "Cells are grouped together under the idea of sculptors or builders because they alone in reality build the organisms."

As to this fact, I agree with him, but I do not agree with him when he states that the cause of the building is only a chemical and mechanical force, and that it all comes about by chance. The inorganic forces like electricity, heat, etc. do not produce anything for a purpose nor do rocks and dead matter produce machines or houses. Mr. Haeckel states: "We notice the very remarkable fact that the egg cell in its original condition is so exactly the same in man as in all other animals that it is impossible to discover any essential difference."

He gives the following figure 9 to illustrate this, and then he goes on and gives a description of the cell to show what a perfect and complete animal he is. He says: "In order to be thoroughly convinced that every cell is an independent organism, it is only necessary to trace the active phenomenon and development of one of these tiny bodies. We then see that it performs all the essential life functions which the entire organism accomplishes. Every one of these little beings grows and feeds itself independently. It assimilates juices from without, absorbing them from the surrounding fluid. The naked cells can even take up solid particles at any point of their surface and therefore eat without using any mouth or stomach. Each separate cell is also able to reproduce itself and to increase. This increase generally takes place by simple division. The nucleus parting first by a contraction round its circumference into two parts, after which the protoplasm likewise separates into two divisions. The single cell is also able to move and creep about if it has room for free motion and if not prevented by a solid covering. From its outer surface it sends out and draws back again finger like processes, thereby modifying its form. Finally the young cell has feeling and is more or less sensitive. It performs certain movements on the application of chemical and mechanical irritance. Thus we can trace in the single cell all the essential functions, the sum of which constitute the idea of life, feeling, motion, nutrition

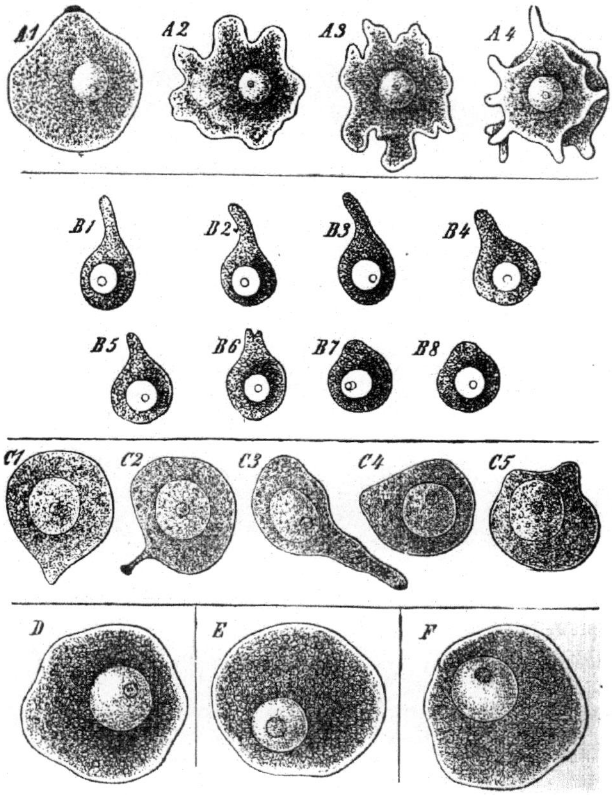

FIG. 9.—Primitive eggs of various animals, performing amœboid movements (very much enlarged). All primitive eggs are naked cells, capable of change of form. Within the dark, finely granulated protoplasm (egg-yelk) lies a large vesicular kernel (the germ-vesicle), and in the latter is a nucleolus (germ-spot); in the nucleolus a germ-point (nucleolinus) is often visible. Fig. *A* 1—*A* 4. The primitive egg of a Chalk Sponge *(Leuculmis echinus)*, in four consecutive conditions of motion. Fig. *B* 1—*B* 8. The primitive egg of a Hermit-crab *(Chondracanthus cornutus)*, in eight consecutive conditions of motion (after E. van Beneden). Fig. *C* 1—*C* 5. Primitive egg of a Cat, in four different conditions of motion (after Pfluger). Fig. *D*. Primitive egg of a Trout. Fig. *E*. Primitive egg of a Hen. Fig. *F*. Primitive human egg.—HAECKEL.

and reproduction. All these properties, which the multicellular highly developed animal possesses, appear in each separate cell at least in youth. There is no longer any doubt about this fact and we may therefore regard it as the basis of our physiological idea of the elementary organisms."

After this general description of the cells of the different parts of the body, he makes the following statement in reference to the brain cells: "In the protozoa in the one cell plants and primitive animals, the whole organism permanently consists only of a single cell. On the contrary in most animals and plants, it is only in the earliest

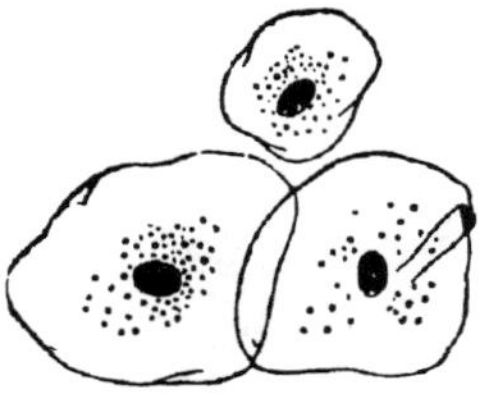

FIG. 10.—Three epithelial cells from the mucous membrane of the tongue.—HAECKEL.

period of individual existence that the organism is a simple cell. It afterwards forms a cell society or more correctly an organized cell state. The human body is not in reality a simple life unit as is at first the universally current simple belief of man. It is rather an extremely complex social community of innumerable microscopic organisms, a colony or a state consisting of countless independent life units of different kinds of cells. * * * All the numerous tissues of the animal body such as the entirely dissimilar tissues of the nerves, muscles, bones, outer skin, mucous skin and of other similar parts are originally composed of cells, and the same is true of all the various tissues of the vegetable body. These cells, which we shall hereafter consider more closely, are independent living beings, the citizens of the state, which constitute the entire multicellular organisms. We thus have before us a highly complex apparatus, the more minute structure of which we have hardly begun to know, even with the help of our strongest microscope and the significance of which we rather guess than know. Its complex mechanism is capable of the most intricate psychical functions. But even this elementary organ of mental activity of which there are thousands in our brain, is only a single cell. Our whole intellectual life is but the sum of the results of the activity of all such nerve cells or mind cells."

The living structures that we see include all those things containing life such as plants, trees, insects, animals, birds, fish, etc. They are all produced by the cell. They are constantly changing, a fact which proves that the builders are all the time trying to improve their habitation in order to meet some climatic or other condition, or obtain some advantage in one way or another in the struggle for existence. By reason of this fact, the scientists have decided that it is impossible to classify them as animals and plants because they overlap each other so gradually that it is impossible to tell where the one begins and the other ends. We might classify them generally as either movable or as stationary structures. The plant building cells usually build stationary structures because they are able to make their own food and building material by the aid of sunlight from the raw material of earth, air and water. The animal building cells on the other hand being compelled to move about in search of food build movable structures. The classification would not in all cases be correct, as some plants have movable parts such as sensitive and insect catching plants. However, as a general classification, it is as good as any to consider them as either stationary or movable habitations of the cell, just as a house is a stationary habitation of man while a ship is a movable one.

Some of the simplest structures are mere associations or groups of cells like the micro gromia socialias. These cells understand how to inclose themselves in shells, and stick the shells together, and in that way remain together in a social community. They have already discovered the art of building covers or armours about themselves of

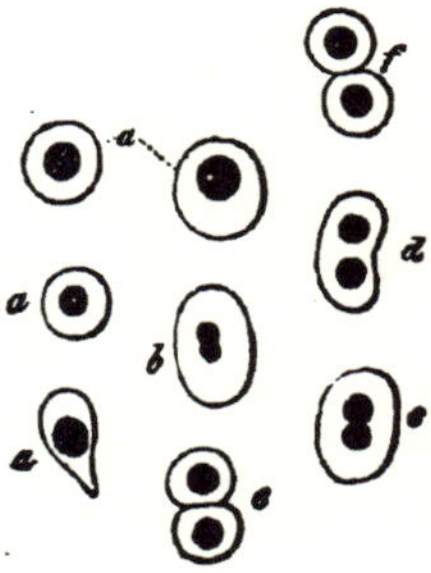

FIG. 11.—Blood-cells, which increase by division, from the embryo of a young stag. Each blood-cell has originally a kernel, and is globular *(a)*. When they are about to increase, the cell-kernel, or nucleus, first separates into two kernels *(b, c, d)*. The protoplasmic body then becomes pinched in at a point between the two kernels, which become more widely separated from each other *(e)*. Finally a complete separation between the two parts is effected at the point where the original cell was pinched in, so that there are now two cells *(f)*.—HAECKEL.

lime and other materials, and the advantage of associating together for offensive and defensive purposes in the struggle for existence. Fig. 12 is a description of this animal taken from a text book on zoology. This cell is similar to the cell heretofore considered under the name of volvox. Man in his first days of development is also a mere cluster of cells, the same as this. After the cluster is formed, then these arrange themselves into layers in the shape of a cup, which is the beginning of the stomach. A wonderful plant or animal, whatever a person may wish to call it, is the Physalia. It is also called the Portuguese man-of-war. In a book on Zoology it is briefly described as follows:

"One of the most remarkable and best known of this group. It consists of a pear shaped and elegantly crested air sac floating lightly upon the surface of the water and giving off from its under surface numerous long and varied appendages. These appendages are the different members of the community and perform different functions,—some eating for the whole, others producing medusi buds

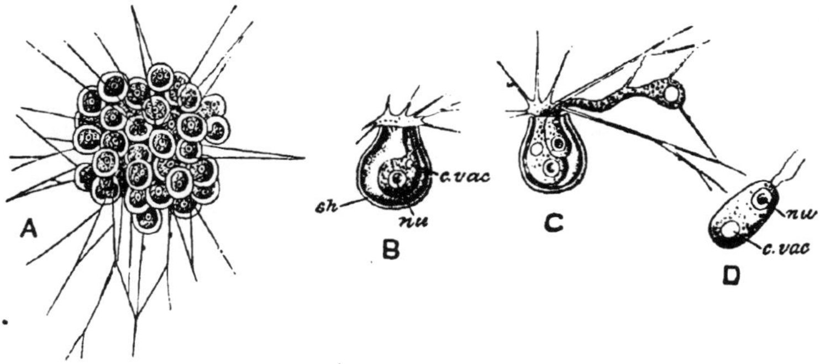

FIG. 12.—*Microgromia socialis.* *A,* entire colony; *B,* single zooid; *C,* has undergone binary fission, with one of the daughter-cells creeping out of the shell; *D,* flagellula; *c. vac.* contractile vacuole; *nu.* nucleus; *sh.* shell.

and others being the locomotive members, the latter having tentacles with powerful stingers that stretch out behind the floating community. The air sac is three or four inches long."

The social community of cells that build this structure called Portuguese man-of-war, show wonderful skill, as they are able to sink or swim at will. The Albatross Expedition describes them as follows: "The slime on the sea weed for instance may have come from a most poisonous jellyfish known as the Portuguese man-of-war, that

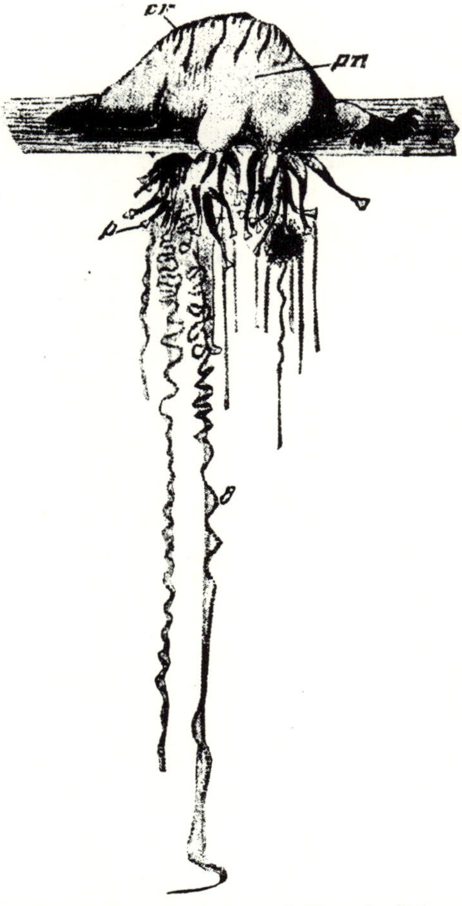

FIG. 13.—*Physalia* or *Portuguese Man of War;* the living animal floating on the surface of the sea. *cr.* crest; *p.* polype; *pn.* pneumatophore.

has stung to death many valiant swimmers with its ten foot streamers, that paralyze the body. On fair days these formidable creatures lift their pink and blue oval crests above the surface, but in rough weather they sink and one of their streamers entangled in a tow net might do serious harm, as happens on the slime banks of Behring Sea, where the cod fishermen suffer grievously when their lines come up smeared with a poisonous jelly fish excretion. It was the 'Albatross' that traced the cause of this Behring Sea fishing trouble to a jelly fish."

We see that this colony of cells can sink when necessary to evade the crushing force of the waves and again float when the weather and conditions are favorable. This animal looks like the ordinary sea weed floating on the water but upon a closer investigation we find that this is more than a mere social gathering of the cells. Here they have discovered the advantage of specialization; each bunch of cells takes charge of its particular line of work, such as moving about, reproduction, capturing and devouring smaller individuals for food. You notice that they have also discovered a method by which they kill and capture other animals for food by a powerful sting. It is not likely that they can see or hear in the sense that we understand that power. They can, however, feel, smell and taste. The thinking capacity of the cells that make up this social community must be the best, when we consider the discoveries and inventions they have perfected in the course of their development from the single cell to the present state. Consider the million of individuals involved and occupied with their different kinds of work, in the make up of this social community. This colony of cells we call Physalia or Portuguese man-of-war is called an individual.

We might next consider the Star Fish. It has one eye; and has feet with sucker ends so that it is able to walk almost anywhere and in any direction. It has developed a very good smelling apparatus and is very well equipped for life. It is encasèd in a star shaped structure, strong and tough as bone and its star shape makes it a disagreeable substance to swallow by any larger animal. The cells that build this structure have been able to hold their own in the struggle for existence and they have reason to be proud of the progress they have made in the past ages.

We might consider next the species of sea plant or ani-

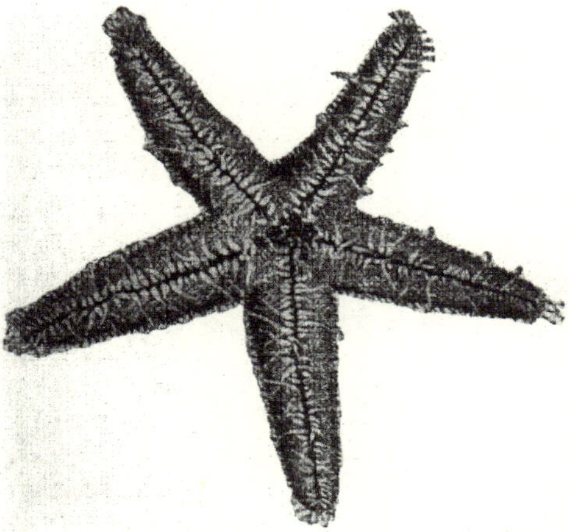

FIG. 14.—*Starfish.* General view of the ventral surface, showing the tube-feet.

mal called the antedon. This animal fastens itself on the sea bottom. It starts to grow in the same manner as any other animal. The single individual cell who starts the building of this animal, first swims about in the ocean, then finally settles down and builds this structure. The secret of the success and progress of this individual is in the fact that it covers itself with such disagreeable building material that other animals will have nothing to do with it and cannot use it as food. This animal has been able to produce electric lights in different places of its body, as you will notice from the following by the Albatross Expedition: "In the midst of gorgeous submarine forests and waving gardens that fringe the reefs of the ocean floor and spread over its vast plains are abundant clusters of shining trees or bushes, known as sea feathers or sea pens, these also being animals, not vegetables.

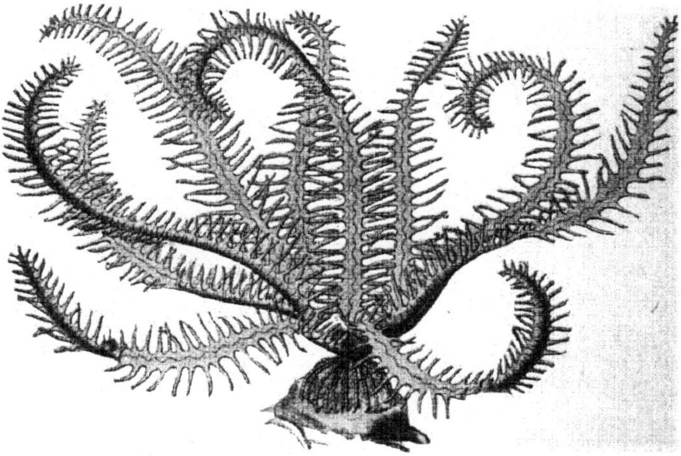

FIG. 15.—*Antedon*. Side view of entire animal.

Their long stems glow with a dull phosphorescent light when the trowl nets bring them up from the depths, and if they are touched with ammonia they shine brightly. It is thought that the light is dulled through their fright in capture and it is probable that normally they give forth a brilliant radiance when they desire to attract their prey or to terrify their enemies. * * * Never was there an animal so lacking in any immediate usefulness as the

Crinoid. It cannot move, it has no eyes; it makes no attack; it does no harm. It simply eats, playing the part of universal scavenger of the seas, catching all foods that fall through the waters, animal and vegetable, in its ten or more waving arms, each of which has a long groove lined with propulsive hairs that work the food along in the manner of a moving stairway to a central mouth and stomach. This stomach lies between the basis of the arms, which rest either upon a long stalk or upon two or three dozen legs, that cling fast to rocks or other animals or spread out upon the surface of the mud. The crinoid is perhaps the only creature in the sea that is not desired as food by some other creature, but these animal lilies, which eat everything, are not themselves to be eaten, being too brittle, too full of lime, all skeleton, as it were. Even the stomach of a crinoid has its own skeleton."

You will notice that this animal is just as much a plant as an animal. There is no difference in the beginning nor in the development of a plant or animal. It is put up for a purpose. If it is to be a stationary structure, it will look like a plant; if it is to be a movable structure it will look like an animal. It all depends upon the purpose for which it is made. This animal, which grows like a plant in the bottom of the sea, is merely a house lit up with phosphorescent lights wherein dwell millions of individual cells.

We might next consider a very large group of structures known as shell animals. The cells live in a social colony which is protected with a shell made of lime and other material. The shells that protect them are usually of such strength and thickness as would be required to withstand the pressure and force of the water at that particular place, or the animals that may attack them. They are simply movable houses wherein live vast colonies of cells. These cell colonies are generally specialized in their work, as some attend to locomotion, some to digestion, etc. Many have also developed special sense organs like eyes and ears. They are all well adapted to their

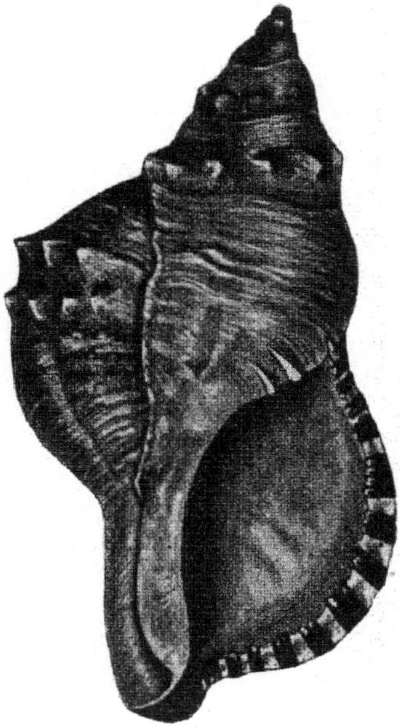

FIG. 16.—Shell of *Triton nodiferus.*

place and condition in life and they live secure and comfortable in their strong and beautiful habitations.

We might next consider the individual known as the cuttle fish and he is a wonder. The cuttle fish is covered with a hard shell made of carbonate of lime and other hard material. Its size is up to three feet in length, so it is small enough to have a number of enemies. It is pro-

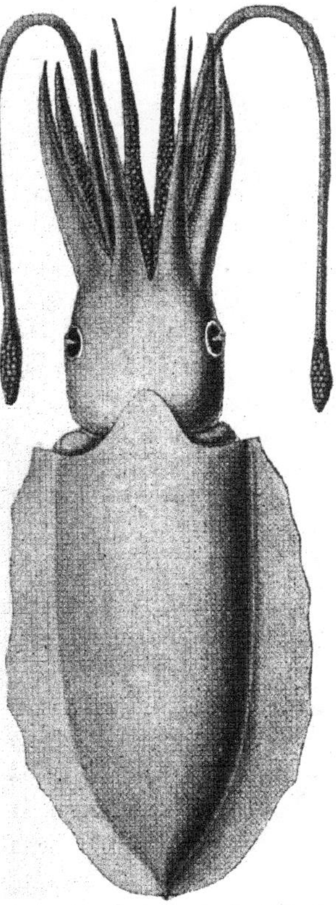

FIG. 17.—*Sepia cultrata* or *cuttle-fish*. Entire animal viewed from the dorsal aspect.

vided with a sack of black material like ink, which it throws into the water behind it when pursued by enemies and in this way it escapes very easily. It has long tentacles provided with powerful suckers, which hang on to any animal that touches them. It does not look like an animal, and for that season it is able to get near fish and other animals on which it feeds. The scheme of throwing ink in the face of its pursuers is not the only wonderful invention of the cells that build this individual. The most wonderful is its power to chaange its color at will. If its surroundings are green, it will change from black to green, etc. How is it able to do this? It has been demonstrated that the cells in the skin get a picture of the outside surroundings through the eye, that the skin cells have several color shades on hand and that they stick those tints up to the surface of the skin, which will produce the shade desired. It shows how the cells in the individual communicate and work with each other. The skin cells gets a picture of the outside from the brain cells, who must first get it through the eye. If you injure his eye or the nerve leading to the eye, he is not able to change his color. This invention of changing his color at will so as to be able to escape enemies or get closer to its victim is certainly wonderful. How long the cells struggled with the elements and enemies before they discovered and perfected these inventions to aid them in the battle of life we can only guess; but from the time it first discovered the advantages of working together in a social community until it perfected the eye, next the ink scheme and the power to change color, it must have been ages. The eye must have been perfected first, or else it could not know the effect of roiling the water with ink, or of changing its color to correspond to surroundings.

We might consider now this wonderful structure known as the torpedo ray, or sting ray, as it is sometimes called. This is a fish that kills its victims with an electric shock. Here is a colony of cells that have built a habitation with which to move through the water like a fish and have also built up in it electric storage batteries,

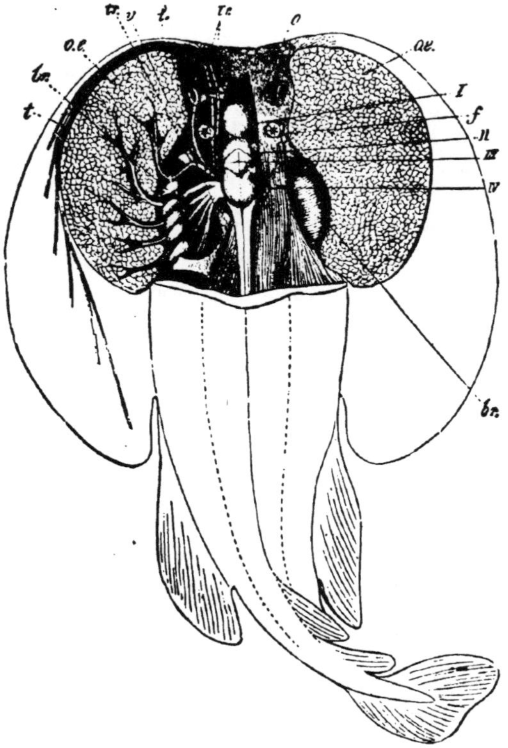

Fig. 18.— A *Torpedo-Ray* with the electric organs dissected out. On the right the surface only of the electric organ *(oe)* is shown. On the left the nerves passing to the organ are dissected out. The roof of the skull is removed to bring the brain into view. *br.* branchiæ *f,* spiracle; *o,* eyes; *tr,* trigeminal; *tr'*, its electric branch; *v.* vagus; *I,* fore-brain; *II,* mid-brain; *III,* cerebellum; *IV,* electric lobe.

in which it is able to collect and keep enough of electricity to give a powerful electric shock to its prey or enemies. Think of all the experimenting that must have been carried on, until they perfected a machine with which to gather electricity from the surroundings and keep it stored in batteries ready for use. It could not have been of any advantage or use until they had it perfected and all the machinery working. Until that time they must have had a committee of cells continually working on the idea. They could not have commenced experimenting until they had previously discovered the effect of the electric shock and could see the idea of using and developing it into a weapon for offensive and defensive purposes.

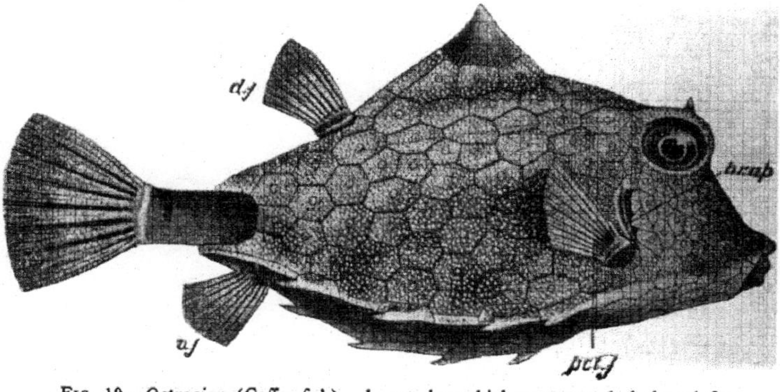

FIG. 19.—*Ostracion* (Coffer-fish). *br. ap.* branchial aperture; *d. f.* dorsal fin; *pct. f.* pectoral fin; *v. f.* ventral fin.

Then we have the coffer fish. This individual is a perfect box made of some horn like material in nearly the shape of a fish. The box has an opening for just the necessary steering and propelling apparatus and the eyes and mouth. It is protected with a color conforming to its surroundings but it cannot change its color. It is a wonderful submarine boat, well protected with armor and so constructed that a larger fish will not eat it unless very hungry, as its spikes make it a very disagreeable morsel to swallow. The strong covering and protective color give it such advantages that it has been able to hold its own with others in the struggle for existence in the past, and is with us today and not numbered among the extinct species.

Another most wonderful individual is known as the Stomiasboa or lantern fish. This fish is black with phosphorescent shining light in front, and two rows of bull's

FIG. 20—*Stomias boa or Lantern Fish.* The white dots are the luminous organs.

eye lanterns the whole length of the underside of his body. These spots all shine like little electric lights and the black form of the fish cannot be seen in the night, but only his spots, which also light up his surroundings, so that he can see his victims with his extraordinarily developed eyes adapted to see in the dark. He is covered with a powerful scale, adapted to resist the pressure of the water two miles deep in the ocean, where he stays in the day time and where his enemies, the larger fish, can not follow him. Think of the wonderful skill exhibited in the construction of this individual,—his powerful eyes and mouth, with his powerful search light at the end of his nose enabling him to see in the dark. These things are all necessary to an existence two miles down in the water. This perfect arrangement of everything to meet conditions, exhibited in this individual, must have also required ages of experimenting by the best intellects. The Albatross Expedition describes him as follows:

"Among the strangest of these ascending night feeders are the lantern fish, remarkable for this, that their bodies are dotted over with electric lights; certain round phosphorescent spots arranged in rows along the sides, that glow brilliantly just as fire flies glow, especially a large spot on the end of their noses that shines like a search light. So these queer fish move through the water, ascending and descending—small submarines all ablaze. There may be a double usefulness in these phosphorescent lights which flare up suddenly against a deep sea enemy and frighten him away, or which lure the prey at higher levels, as a candle lures the moth."

It must be conceded that to be able to build and maintain a submarine like this we call the lantern fish, which can light up its surroundings and adjust the structure to resist the pressure of the water at different depths up to two miles, is a task requiring the best engineering and keenest intellect. Think of the details to be looked after to keep it adjusted to the ever changing environment, to keep all the lights going, to gather the material and manufacture the light.

Fig. 21, the Lure Fish, is another illustration of a fish that tempts its prey by means of a phosphorescent light. He has a bait that shines, attached to the other end of the string which he lets out like a fishline to attract curious individuals. The string to which his bait is attached is such that he can let it out and pull it in at will. This you will see is a moving habitation something like a submarine with a fishing apparatus attached.

We never stop to consider the position occupied by the individual cells that build animals and plants, who must conceive and execute ideas. While this submarine like

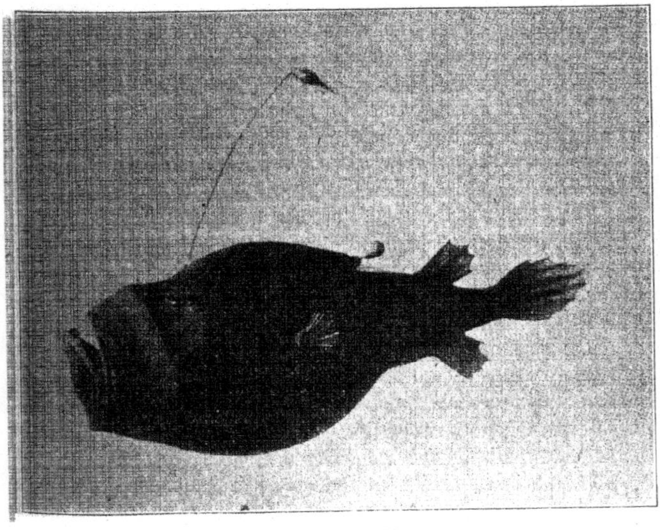

FIG. 21.—*Lure Fish.*

individual was lying in wait for his victim, it must have occurred to the cells occupying and in charge of it that animals are attracted by a light. That if they could fix up a shining bait, which they could let out a short distance to attract attention, they could hide their submarine and decoy their victims within reach of their grappling apparatus, called the mouth, arranged in front of the individual. From this idea conceived, they would proceed to build this fishing apparatus. This is what the Albastross Expedition has to say about him:

"The potency of light in attracting wanderers of the deep is seen in the equipment of the Lure fish, a grotesque creature with a huge mouth that hides its black body in the mud, and waits patiently for victims, dangling before them a phosphoresent bulb that shines at the end of a long filament—a self-grown and self-baiting fishing rod, curving forward from the animal's head and hanging temptingly before its hungry jaws, ready to snap open at the approach of a curious visitor. These Lure fish are found at the depth of three miles or more."

Mr. Muffit gives the following interesting description of a few of the enormous number of different kinds of individuals found in the ocean:

"Another danger lurking in the tow nets is the possible presence of a strange crustacean, related to a crab, an uncanny creature about three inches long that is invisible, literally invisible, owing to the fact that its head and body, its arms, legs and claws are quite transparent. The presence of this animal in the receiving pan is usually indicated by a disturbance among its visible neighbors, the shrimps and fishes, and when it is lifted out with a pair of tongs it appears like the glass model of a crab with slowly moving glass legs and glass claws. When killed this crab loses its transparency and reveals itself in a dull white coloring like the white of an egg. The 'Albatross' encountered many of these invisible wrigglers while fishing in Japanese waters. It is well known that very young fish and tiny eels are quite transparent except for two black dots, which mark their eyes.

"Each haul of nets brings up some deep sea wonder.

It may be the hideous viper fish with teeth so long that they fold outside of his mouth like the tusks of a wild boar, or the snipe eel with its bill like its name sake and a body like a length of whip cord or the queer pelican fish that will swallow a fish much larger than itself and somehow digest it, or a dead ribbon fish with its almost transparent body, 20 feet long and a foot wide and half an inch thick, or a great red jelly fish full of poisonous darts coiled up in its body, and ready to shoot out their venom against any touch. In New England and arctic waters some of these jelly fish grow to enormous size, their bodies measuring six or eight feet across and their pendent streamers reaching down seventy feet or more. I may mention also the giant squid or cuttle fish thirty feet long, a whitish colored beast that is always found dead, the same being true of the giant octopus with its reach of seventy feet from tip to tip of its huge arms.

"The jewel beauties, swimming about rather tamely would be helpless against the ravenous pursuers were it not that they live in shallow tide pools and near coral reefs where these pursuers dare not follow them. Why not? Because coral reefs are full of stings of the live coral creatures, stings that hurt a man's hand if he touches them and might destroy the eyes of any big fish that ventured among them. And tide pools abound in sea urchins with sharp barbed spines, hundreds of them that break off inside the wound." The fierce struggle for existence that has been going on in the sea for ages has produced these wonderful individuals, in the same manner that the struggle for place and power that is now going on in the war between the English and the Germans is causing inventions of many kinds, and weapons and engines of destruction of different characters. We must keep in mind that the builders of all these individuals are the cells, just as the builders of all the different structures now used in war in their struggle for national existence, are the human individuals. The structures produced by cells or cell communities are all on the plan intended to resist attacks by hungry enemies; take, for instance, the common turtle with its armor plate of bone, which in every way resembles a fortification. Millions of living beings working together live inside of this moving fort we call the turtle.

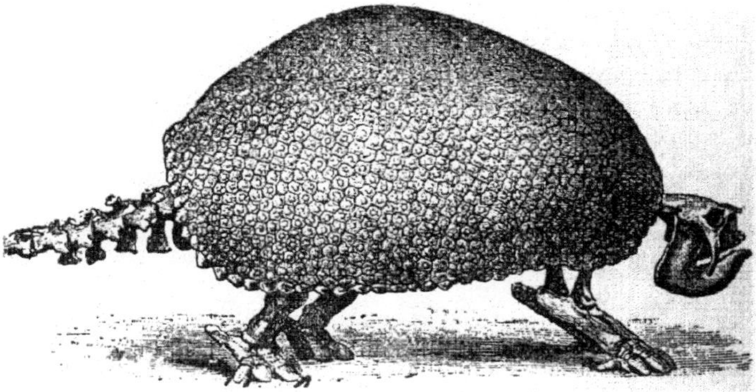

FIG. 22.—Skeleton of Turtle. *Glyptodon clavipes.*

What difference can there be in the intelligence or skill required in building a submarine by human beings and the building of a fish by the cells? In every case there must be a preconceived plan, a purpose to capture and escape, to do this or that. They are made with an end and purpose in view. The time will evidently come when every important battle will be fought either in the air or under water. The submarine now must have eyes in order to see where to go and so as not to run into nets. Inside are the individuals who run it and take care of it precisely in the same manner that the cells run and take care of the individual they inhabit, be it animal or plant. Think of that invisible crab! How are the cells able to build themselves into a structure so as to be transparent? Those are secrets for us to solve.

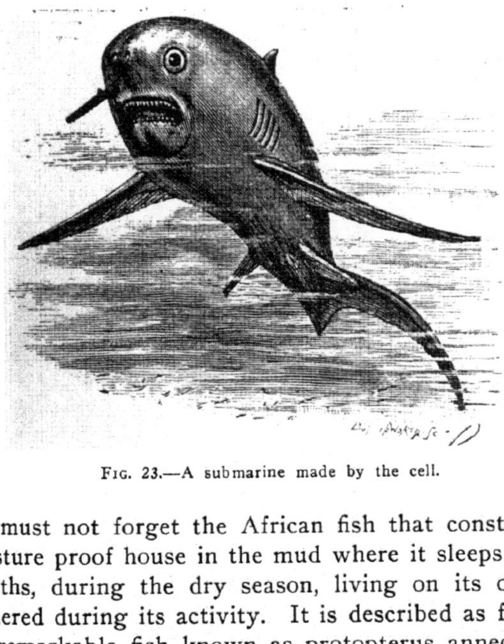

FIG. 23.—A submarine made by the cell.

I must not forget the African fish that constructs a moisture proof house in the mud where it sleeps for six months, during the dry season, living on its own fat gathered during its activity. It is described as follows: "A remarkable fish known as protopterus annectens is found throughout the whole of tropical Africa, but is most common near the West coast, where it sometimes attains a length of six feet. During the dry season, when many of the ponds dry up, the fish descends some distance into the mud and forms a rounded hollow for a

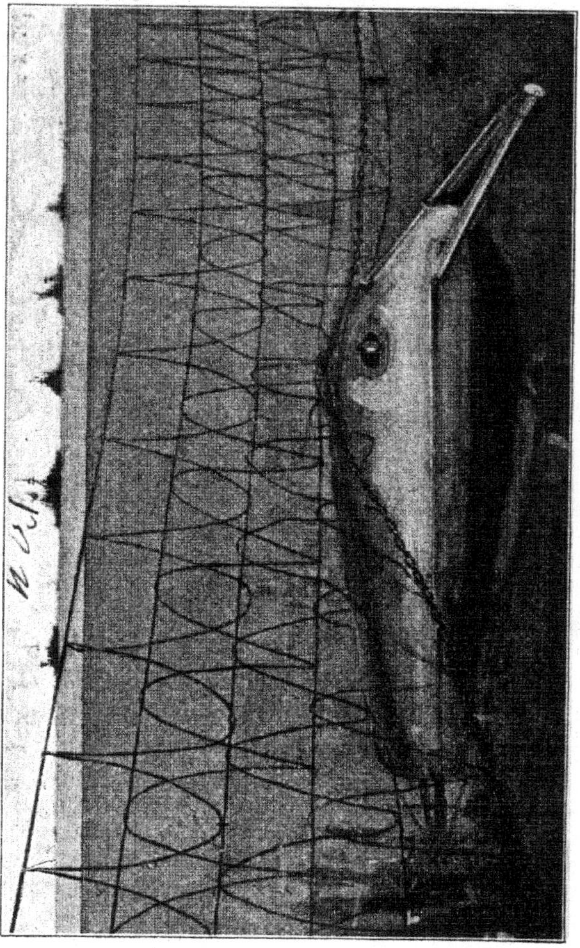

Fig. 24.—A submarine made by man.

nest, which is lined by a capsule of hardened mucus secreted by the glands of the skin. It hibernates thus for nearly six months, drawing its sustenance from the fat secreted when it is active." This fish is wonderful in that it is able to make a nest in the mud, where it will not lose any of its moisture, by lining its room with a secretion which will prevent the escape of the moisture. During its activity the cells in charge of this individual gather and store away enough food in the way of fat so they can live and enjoy life until the rainy season comes again and provides the water to float them. The ability of animals to store away food to be used at a future time is a common habit, that is, with all animals that sleep in the winter, like badgers, bears, etc. The cells in the body gather and store food for the purpose of tiding over a time when food will be scarce and hard to get. What possible difference can there be in the intelligent purpose evidenced by the cell in providing for the future, and the animal and man doing the same thing? There can certainly be none whatever.

Take again for instance protective coloration. The military experts are now adopting and taking advantage of the tricks of the cell. The zebra in its bright stripes is almost invisible in the jungles of his natural habitat. When the cells building the zebra gave him the stripes which make him so conspicuous in the circus they knew what they were doing. The experience of hunters all testify to the fact that the tall grasses and trees in the jungle where he lives make the zebra almost invisible. In reference to the cause of the color of animals, plants and fish we have now fully demonstrated that it is caused by the action of cells occupying the individuals, especially if they have eyes with which to take a picture of the outside surroundings, and transfer this to the skin cells, who are in charge of the color business. The following is a very interesting article on this matter, altho it has been demonstrated several times before, and in one case the

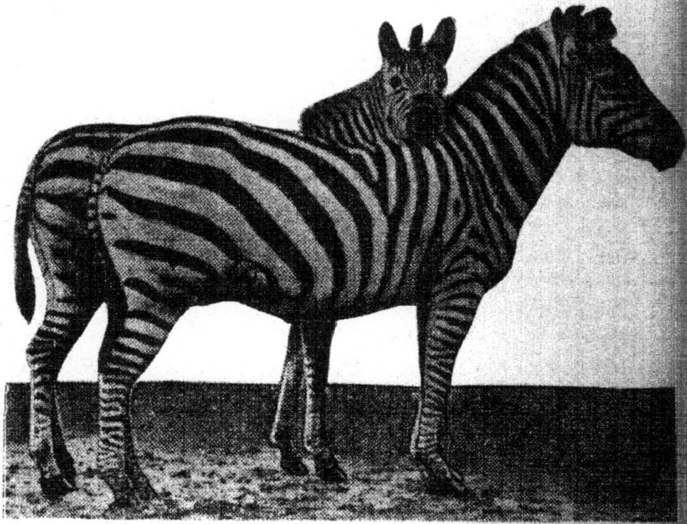

FIG. 25.—Zebra.

fish was able to produce a checker-board on his back when the bottom of his aquarium was painted in that way. S. O. Mast of Johns Hopkins University has published in the "Proceedings" of the National Academy of

Sciences, April, 1915, some results of observation and experiments made at the U. S. Biological Station at Beaufort, N. C. The work is to be published in full in the bulletin of the U. S. Bureau of Fisheries. He says:

"Nearly all fishes simulate their environment to some extent and some flounders do so with remarkable accuracy and rapidity. It was found that flounders in glass dishes became nearly white on a white ground and nearly black on a black ground. They also assumed approximately the colors of all grounds, except red. Fine and coarse patterns in the ground produced correspondingly fine and coarse patterns in the skin, but there was no actual reproduction of patterns. Five days' sojourn in a black pan was required to produce a maximum blackness in a flounder that had been kept two weeks in a white pan, but the change from white to black was effected in two minutes in the same flounder after it had been transferred from one pan to the other. The change from black to white always required an hour or more. Color changes are comparatively slow. Yellow usually predominates in the environment and is assumed more rapidly than green or blue.

"The skin of flounders contains black and yellow cells called chromatophores and opaque white cells called iridoeytes. The changes in color and pattern are produced by changes in the arrangement of the colored cells and in the extent to which they are hidden by the white cells. These changes are regulated by ocular impressions. Flounders become uniformly white when the head-end is placed on white and the tail-end on black. They become black when the head is on black and the tail on white and they become gray when one eye is on white and the other on black. Exposure of one eye to a fine and the other to a coarse pattern produces a combination of a fine and coarse pattern in the skin. The influence of each eye extends over the entire body. The skin becomes yellow when a yellow card is placed very near the head. A flounder deprived of one eye simulates the background quite normally, but there is no simulation whatever when both eyes are removed. Flounders fail to simulate the ground in very strong illumination from above and they become white on all ground when their eyes receive no light directly from above. Adaptation to the ground is not effected by covering the skin with sand, so that the fish cannot see it. Vision in fishes is very like human vision in regard to shade and color, but less acute in regard to size. Flounders distinguish between dots of two millimeters and three millimeters and recognize dots of one millimeter, but not those of five-tenths millimeters. By means of a rotating background of black and white sectors the acuteness of vision in regard to motion was found equal to that of man. Flounders adapted to a given color seek ground of that color, and color in the skin is produced only by exposure to the same color."

It is a singular thing that scientists and the human mind by reason of vanity or for some reason will not conceive, or are not able to see that the minds that direct these very difficult acts of effecting these protective colors in fishes and animals are the cell minds and cell intellects. They build the fish and take care of it for their own selfish purpose. They alone are responsible for its success. These are facts we do not have to guess at. We do not have to spin any theory about it because we can see it. In a daily paper I read the following in regard to the acts of man and animals in reference to this same trait of taking advantage of protective color: "If man has learned clever tricks for deceiving his enemy he has been taught by nature. In all nature there is scarcely an animal which is not characteristically marked for deceiving his enemy. Many butterflies are veined and marked like leaves and flowers with such splendid accuracy that when they are hidden in the petals of the flowers or hovering on the foliage they are not to be distinguished. Likewise the walking stick insect when it crouches among the green leaves cannot be differentiated from the twigs. Bird hunters are aware that it is difficult to discover grouse and partridges because of their bark-like coloring.

Nature is so cautious in trying to protect many of her children against their enemies that she often changes their colors with the seasons. Most of the arctic animals change from brown to white as the winter approaches. This is no less true of many kinds of fish. Some have the facilities for harmonizing colors with the particular character of bottom upon which they happen to be resting at the moment.

The lesson of nature has not been lost upon man. From time immemorial bird hunters have clad themselves in green, that the animals would not distinguish them from the foliage and the surroundings. The American soldiers were the first to adopt the khaki uniform, because it is the color of the earth. European armies have now acted on our suggestion, although it has taken a long time to teach the French soldiers that the bright red on their uniforms is a menace, not an aid.

Just at present news dispatches tell of the increased number of changes in the army uniforms in an effort to make the soldier invisible—or as nearly so as the ingenuity of man can make them. Russian artillerymen and scouts have been clad in long white cloaks and caps while fighting in the snow fields of the Carpathians.

It is said that the adoption of nature's methods of protecting animals has been so successful that it is impossible to distinguish moving men from waving shrubbery when but a few yards distant.

Practically every army involved—and they include many nationalities and many picturesque styles of uniform practically unheard of on this side of the Atlantic—is adopting this method of uniforming, to some extent.

You see that these acts of man and animals, which we consider very intelligent when performed by man as an individual, we refuse to consider intelligent when performed by animals or more correctly speaking by the animals we call cells, who build the animals and perform the work.

There is a moth called, "The Death's Head Moth," because it has the resemblance of a skull and cross-bones on its head, which is merely an arrangement as a protective color. A scientific magazine makes the following remarks about it: "The Death's Head Moth not only has a mouth with which to eat, but it can make a noise which resembles that of a mouse. It is the only moth which makes any sound. It is this peculiar sound which it makes, as well as its resemblance to a skull and cross-bones marked upon its head, which makes superstitious people afraid of it, for they believe that it brings them trouble. The moth though forbidding in appearance is entirely harmless, of course."

Now who are the most intelligent beings, the brain cells directing the action of the man afraid of this moth, or the cells that build and direct the moth? Which of the two are the most profound thinkers? It would seem that the cells in the moth discovered the superstitious nature of man and adopted this style of protective color to frighten him.

In reference to the intelligence of insect-building cells of all kinds many of which are also able to change their color at will, space will not permit going into details. In their place in life, the struggle for existence is fierce and competition very strong, and they have also discovered nearly every trick and method used by fishes and animals.

The following article by A. L. Hodges states the situation in a general way: "Few people are familiar with the fact that the diving bell was invented by a spider. Such however is the case and if it was not actually invented by him it was certainly used by him long before our hydraulic engineers made one for the same purpose. The diving bell is, as is well known, a cup-shaped body with open end down which is let into the water. The air is caught in the bell and keeps the water from rising beyond a certain level at any specified depth and of course allowing anyone inside to breathe and act as if he were on the ground. The new improvement of the diving bell, known as the Caisson, is a huge pipe which has compartments into which the air is pumped from above. The spider's bell is filled more in this manner than in the other.

"The name given to these little spiders is very appropriate—Naiads of the family of Arachnida. The Naiad will build a little house of water-proof silk, held fast by strands fixed to neighboring blades of grass and stones, several feet under water. He completes the entire structure before filling it with air—as if he knew that the air would tend to make it rise to the top and thus hinder the attaching of the anchors.

"But the method of getting air into their houses is perhaps the most peculiar and interesting of all instinctive acts of animals. Their abdomens are so made that a bubble of air can be caught underneath them. This the Naiad does, and swims to his house with it and turns it loose in the airy structure. The process is repeated several times until the little house is full of air. Of course the open end of this house is down and this has to act also as the entrance to it. In the little water-house the spider spends the winter and rears its young ones. The house also acts as a lair from which the spider can jump on unsuspecting prey.

"Another peculiar thing about the Naiads is that they never get wet. They have thousands of small hairs on their bodies which hold and keep the air from being washed off when they enter water and so the air sticks and water cannot approach.

Scientists are acquainted with many other insect engineers, but with none that approach Naiads in intelligence and skill. The Water Beetle is probably the only other one in their class. It builds a water-proof nest under water, but does not live in it. It merely lays its eggs in the nest, seals it up and leaves.

The Mason Bee is as his name implies a builder of structures of stone and mortar. The nest is attached to almost any solid structure and actually does consist of small stone, cemented together with mortar. The house consists of many cells of oval shape, and into each an egg is laid. The cell is lined with silken web by the mother who gets out of it by holding its top. Before leaving, however, she hermetically seals up the cell and leaves the youngster to its fate.

"However, such are the arrangements of nature. As soon as he gets to feeling his oats and consumes the food left him by the mother he finds himself supplied with tools hard and sharp enough to cut through the walls to freedom.

"A member of this family found in England makes his own bricks, selecting brown clay for the purpose, which he mixes with saliva, rolls into small balls which soon become hard and then cements them together. These pellets are as large as small peas and one bee has been known to prepare as many as one hundred and fifty in a single day."

I have watched the action of bees, beetles, spiders and ants and other insects and must say that their actions show intelligence of a very high degree and will refer to many of their acts later on, which are simply wonderful. You might say if the insect building cells are so intelligent, why do they not build larger and stronger structures like ourselves or other animals. Upon that point we need but consider all the structures produced in the past ages, like the Mastodon and thousands of others that went too far in that direction and got the structures too large. Their skeletons are now conclusive evidence to the insects and ourselves that size is not necessarily a quality that spells success. The size must be considered in the light of permanent and perpetual existence on this planet, as well as the other features. I quote the following from the Scientific American, which is very significant: "Rarely is it safe to speak of anything as ultimate in prehistoric life, but there is little doubt that the American Museum now exhibits a skeleton of the largest flesh-eating animal that has ever lived. This is Tyrannosaurus, the tyrant lizard, a dinosaur that lived during the close of the Cretaceous period. It was one of the very last expressions of its race and, judged by size and structure, was king of its kind. An idea of its immense size can be formed from measurements of the skeleton, 47 feet in length, and, as mounted, 18½ feet in height. When fully erect this animal would have reached a height of 20 feet.

"Larger herb-eating dinosaurs have been found in America and East Africa in older rocks of Jurassic or

FIG. 26.—Skeleton of Tyrannosaurus Rex No. 5027, 47 feet long and 18½ feet high.—SCIENTIFIC AMERICAN.

early Cretaceous age, but the flesh-eaters contemporaneous with them were a third smaller than the present animal.

"The Tyrannosaurus was capable of destroying any of

the contemporary creatures and was easily king of the period and monarch of its race."

Now if it is true that the cell is a builder of all living things, both plants and animals, we should be able to find the same intelligent scheme for self-protection and self-perpetuation also in plants, or what we might call their stationary habitations or structures; and such is the case, for the schemes and tricks employed by plants to serve their purpose show intelligence of a high order. It is impossible in this short chapter to go into the details and describe the innumerable methods used by plants to fight drought, animals, frost, heat, etc., but the following article from the Literary Digest is a very good general description of some of the methods. The article intends to describe some of the unnecessary cruelties practised by some plants on animals and insects in order to further their own selfish purpose. It says: "Take the case of the fruit of the Martynia, a South American plant, which is armed with terrific hooks, sometimes as much as five or six inches long, so curved that they seize hold of passing animals and plunge deeply into the flesh. It is said that the Bullocks are often thus driven half frantic and suffer dreadful wounds. Of course the final result is that the seeds receive a very wide distribution, but a large amount of needless suffering seems to be involved. . . . . Even more astonishing is the case of the Grapple Fruit of South Africa (Harpagophytun). This species is of a low growing habit and bears fruits which are freely adorned with most formidable barbed appendages. The fruit secures its dispersion in the following manner: in its position out of the ground it is liable to be trodden on by sheep, deer, etc.; at once, of course the hooks catch hold and these penetrate into the tender places of the foot between the horny portions. The unhappy animals limp

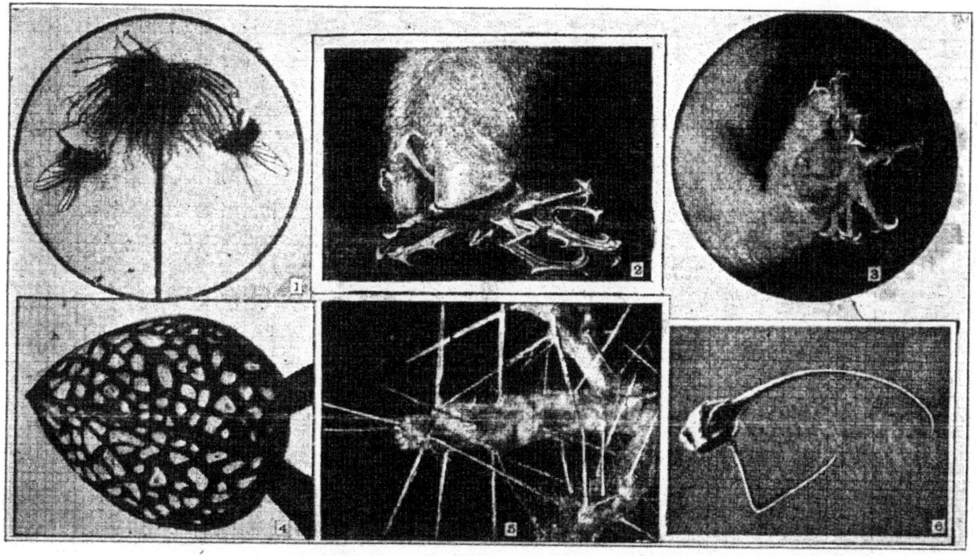

PLANTS THAT COMMIT ATROCITIES ON PEACEFUL ANIMALS AND INSECTS.

FIG. 27.—The Geum (1) captures flies on small sharp hooks from which they cannot escape, although the prey is of no use to the plant. The grapple-plant "attacks" the feet of sheep and dogs (2 and 3), achieving no other end than to cause suffering. The Darlingtonia hood (4) deludes the captive flies with its transparent patches, against which, endeavoring to escape, they beat themselves to death. The terrible spines of the West Indian cactus (5) catch in the flesh and are difficult to extract. The South American Martynia (6) bears upon its fruit enormous hooks which are plunged into the flesh of animals molesting it. "One cannot get away from the idea that most of the suffering involved appears to be quite unnecessary, for there are plenty of instances to show that the same ends can be achieved in less painful ways."—CURRENT OPINION.

about and it may be weeks before the dreadful burden can be thrown aside. During this time the most dreadful wounds are produced and as well the creature is very likely to fall a victim to some beast of prey. In this connection a very singular happening sometimes occurs which is well authenticated; viz., a lion captures an antelope with a grapple root on his foot; when making his meal, the lion gets the hook capsule in his jaws and the barbs speedily become entangled in the mouth parts. The more the lion fidgets the less likely is he to get rid of the encumbrance, but owing to the pain and annoyance the beast cannot leave his mouth alone, so the miserable business goes on. Days pass and the lion is quite unable to eat and as a consequence becomes weak and helpless. So the king of beasts dies, killed by the fruit of the Grapple plant.

It is of course recognized that plants must take certain means to protect themselves against the attacks of animals. Some of the measures which have been adopted are positively vindictive. Take the case of the common stinging nettle. Here the plant is covered with minute hairs which penetrate the skin and at the same time inject an irritant poison, the effect of which lasts for hours. Some of the tropical nettles are much more terrible. The following is an account taken from the Himalyan Journal of Sir Joseph Hooker in which an Indian Nettle is described: "This plant, called 'Mealum-Ma,' attains fifteen feet in height. It has broad glossy leaves and though apparently without stings is held in such a great dread that I had difficulty in getting men to help cut it down. No wonder that the plant is avoided, for if a person is stung by the microscopic hairs the results are appalling. The pain is at first comparatively slight, but after a few hours the effected part feels as if it were being rubbed with a hot iron. Later the most distressing symptoms arise in other parts of the body, which not uncommonly involve the contraction of the muscles of the jaw and other indications which are similar to those to be observed in the case of lock-jaw. In one instance it was nine days before the unhappy individual was free from pain and discomfort. In such a case as this it would seem that a huge amount of unnecessary suffering is involved. It is possible to protect a plant from attack as can be seen in many cases without adopting such brutal methods.

"Self-defense has been carried to a fine art among desert plants, especially the cacti. An array of spines is of course an admirable means of preventing an attack, but many species have carried the matter a good deal further. In some kinds of prickly pear they have minute barbs on their spines and if any animal should even brush up against them the spines hold on firmly when driven into the flesh. They are loosely attached so that the unhappy creature takes away a large number of spines when he withdraws. These remain to produce festering wounds. Another cactus which adds singular hooked spines to the straight variety is called "The Wait-a-bit Plant." The hook holds the clothes or flesh and meanwhile the sharp straight spines do deadly work.

"It is of course well known that a certain number of species find it needful to capture insects in order that they may supplement their supply of nitrogenous food. In most of the schemes the unhappy victim is doomed to undergo the torture of a lingering death. Very rarely is the insect killed at once. First of all let us consider the case of the Darlingtonia, a plant which usually catches winged insects. The flies are lured by honey secretion to enter the hooded process at the top of the pitcher-like leaf. This they do by means of an opening on the underside. The whole of the upper portion of the hood is covered with transparent patches like so many windows. Now when the fly wishes to leave he naturally flies upward toward the light which streams down through these windows. The real opening is hidden in the shade of the under part and passes unnoticed. Thus the flies simply beat themselves to death in a vain endeavor to escape through the transparent places. This proceeding may extend over hours, but it always has one ending. The fly falls exhausted into the fluid at the bottom of the pitcher and is drowned. . . . . .Many flies meet with peculiarly brutal death in connection with the Venus fly-trap. The insect is captured by its legs and held fast; meanwhile it beats its life away in vain endeavors to escape. In conclusion Mr. Bastin says: "The instances given above are only a few out of a very large number which might be brought forward to show that in many ways plants are guilty of great cruelty. One cannot get away from the idea that most of the suffering involved appears to be quite unnecessary, for there are plenty of instances to show that the same ends can be achieved in less painful ways."

This article was originally written by Mr. Bastin in the Scientific American to show some of the unnecessary cruelties practised by some plants on animals and insects in order to protect themselves or to spread their young. However, we have no cause of complaint against the plant for cruel and inhuman treatment, as we never stop to consider the pain or feelings of the plants to further our wishes, so why should the plant pay any attention to ours? I never can forget my astonishment when I first met the drought and animal resisting structure called the "Spined Cactus" found growing on the Western

Plains with spines so arranged that it would be impossible for the deer or buffalo to attack and eat it without suffering terrible torture. Lack of space will prevent a consideration of the birds, except to refer to them in a general way.

The discovery of the art of making feathers as perfect as they are today, must have taken ages. Man was able to perfect a flying machine in a very short time, but he had the experience of the world at his feet to teach and help him. The wonderful skill exhibited by young birds in being able to fly the first time they make the attempt will be explained later. However, it is no more of a mys-

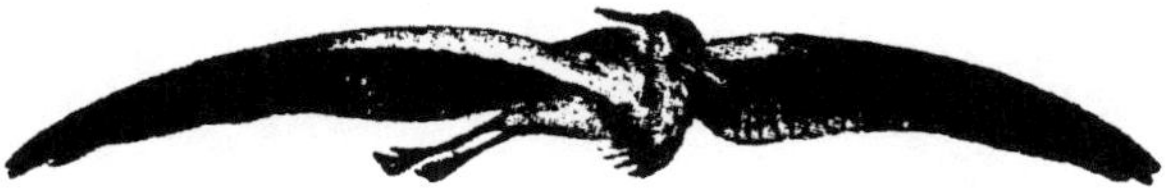

FIG. 28.—Bird descending. (Made by cell.)

tery than the ability of the aeroplane to fly in its first attempt. It is no more difficult to operate a machine than to build it. If the cells understand how to build the aerial structure with which they are able to navigate the air it seems queer that they should not also understand how to operate it. That question will be more fully discussed under the chapter on cause of instinctive action of all kinds, which so far has seemed to be a mystery to mankind. I can see no mystery in the instinctive acts of animals or plants. The builder should know, and does know how to use the machinery he has put together. The living structures which are made by living beings, the cells, are all made for a purpose. For example, the long tongues of woodpeckers and humming birds with which they reach into deep crevices and holes, the web between the toes of the swimming animals and birds, and the long neck of the giraffe with which he reaches up to the high branches of the trees on which he feeds are structures that arise out of the desires and needs of the builders. Take for instance hair, this is a covering mainly to keep the body at an equal temperature. In a cold climate the covering

FIG. 29.—Bird descending. (Made by man.)

must be provided either by the cells in the body or by the acts of the individual himself, as in the case of man. The cells of a man do not make hair on the body, after having discovered a better way of covering the same with the skins of animals or otherwise, which covering can be removed when not required. Hair is produced as it is desired and necessary. On the animals farthest south the hair is the thinnest and on those farthest north it is the thickest, but the evolutionist steps in and states that this is due to natural selection and survival of the fittest,—that is to say, those animals in the north that happen to be born with short hair will be exterminated by the cold weather, and in that way all the short haired animals would be weeded out and only the long haired animals would be left to perpetuate themselves, and that is why we find long haired animals in the north. This proposition you see leaves the whole cause of long hair in the north, to chance and not to desire and intellect. Upon investigation we find, however, that the evolutionist is up against it. You can take for instance, a bunch of pups born in Iowa; keep two in the house all winter, send two to New Orleans and two to Alaska and let two run outdoors and this experiment I believe will settle the whole question. You bring the pups together again in the spring in Iowa and you will be astonished at the difference in the length and thickness of the hair on the dogs. There has been no time to produce the extra length of hair by chance, natural selection or survival of the fittest. Where the climate required the long and thick fur the builders and caretakers of the body provided the covering. The same is true of horses, cattle, cats and other animals. The hair will be short and thin on the animals sent south, and thick and long on those sent north. Everyone knows what long, thick hair the cattle and horses have that have been running out all winter in open sheds, among corn stalks and straw stacks. The necessity of protecting the body from destruction by the cold weather requires long and thick hair. The cells of the surface of the body understand how to get the material with which to build the protective covering, and they increase its thickness and length for that purpose when necessary and not otherwise. The actions of the cells in building hair are brought about precisely in the same manner as the actions of man in making and covering himself with heavy clothing whenever it is necessary and not otherwise. It is his needs and desires and requirements that spur him to action. It is of course true that any person in a cold climate who does not have sense enough to make and cover himself with clothes will perish and be exterminated and this would be true of those animal building cells. The builders of every animal have sense and intelligence enough to increase the length and thickness of the hair if necessary in the same manner that man has, to provide himself with the clothes required. The act requires skill and intelligence sufficient to gather the material and build the structure, be it hair or clothes. One requires just as much intelligence as the other. There is a production of hair for a purpose, just as there is a production of clothes for a purpose, just so we find structures of all kinds produced in the living world for a purpose, as for instance, the shells of the sea animals; some are thick and powerful in order to resist the crush of the water pressure in the deep ocean, or the crash and hammering of the waves on the shore; other shells are thin like those which move over the mud in shallow still water. The shells are the houses in which colonies of cells live. Experience shows that if you plant them in deeper and rougher waters they will build a shell stronger and thicker, just as you might expect, for if the builders act with intelligence in one place they will also do so in other places. This fact has been shown in thousands of ways; for instance, one class of cells will take charge of the work of another class. the cells in charge of the mucus membrane will build an outside skin covering if necessary, or the cells in charge of the outside skin will refuse to build it if it is not the right thing to do. On this point Spencer makes the remark: "That is to say, these literally outer layers of skin are capable of rapidly assuming one another's structures and functions when subject to one another's conditions. Mucus surfaces normally kept covered, become skin-like if exposed to the air, originally moist, tender to the touch and irritated by the air. The surface gradually becomes covered with a thick, dry cuticle and scarcely more sensitive." The facts seem to be that nothing is produced or changed except when it is necessary and for a purpose. It never takes place by chance. We find that differentiation will arise altogether from their method, place and condition of existence; that every plant, insect, bird or animal is a structure designed to meet certain conditions of life, in the same manner exactly as a ship is designed to move over water, an automobile over the ground and the airship through the air. Plants and animals have so many structures in common, which over-lap each other in so many ways, that it is impossible to tell in a great number of cases where the animal commences and the plant ends.

La Mark classified all creation according to the development of the brain and nervous system; such a classification proved later, of course, to be entirely erroneous, as it was discovered later that insects showed intelligence in their particular place in life equal if not superior to any of the higher animals. The cell that builds the little airship known as the lightning bug has the knowledge completely mastered of how to produce a light without heat. Hundreds of thousands of dollars have been spent by the most skilled chemists to learn the secret known by the builders of this bug of how to produce a light without heat. We have finally, after spending considerable money in experimenting, learned from the silk worm how to make artificial silk.

It is impossible to classify the living structures into classes or individuals in a great number of cases, as for instance, in the case of some sea-weeds or animals that grow fastened to the bottom of the sea. At first this creature will be only a single cell, swimming in the water, then it will change and build itself into a fish-like form swiming in the ocean; finally this form will change into a stationary structure settled on the bottom of the sea like sponges or sea-weed.

Every structure, plant or animal, shows clearly that it is made with a purpose of effecting certain ends and that is to satisfy some desire. Take the case for instance of the corn-plant. This plant in order to protect the cob from field mice and other rodents never starts a cob down near the ground. The starchy nutritious kernels embodying the embryo corn must be protected from weather, insects and birds, which is done by a very strong husk. This strong husk prevents the male germs of the plant who are located on top in the tassel from getting to the female germ in the kernel. How did they solve that problem and overcome that difficulty? By building a hollow tube from the female germ in the kernel extending clear outside. When the male germ in the tassel falls down on the silk, which is a hollow tube, leading to the female germ he hunts up the end of the silk and crawls down through this hollow tube which directs him to the place of the female germ. Considered as a whole it is a wonderful scheme and all the difficult problems have been solved and taken care of in the best possible manner. In-breeding must be prevented if possible so the male cells are placed away as far as possible. There is a desire to give the young corn plant a start in life so each cell is provided with food consisting of the starch of the kernel which is sufficient until they have been able to get themselves established in the soil and by the aid of sunlight to make their own food and building material from the raw material at hand. The corn building cells work with a purpose in view arising from a desire to effect certain ends. We build houses, make clothes, produce food. We want and need these articles. We must be protected from the weather. This causes the construction of all kinds of structures to protect us from the elements. The need of food of all kinds caused the railroads and in this manner we can trace every desire to do any particular act to arise from our wants and necessities. The desire has stimulated effort, and effort has devised and conceived structures and methods by which it could be accomplished. In the effort to build a house certain activities take place involving judgment and discretion. Such material will be selected as in the judgment of the builder is most suitable. Every act will involve intelligence, in order that the structure shall conform to the mental picture of what the builder wants. In the same manner does everything that we see come to be and exist from the most complicated city block and railroad system to the smallest living organism. In this particular I agree with Mr. Darwin, who states:

"That animals have a capacity to be modified by processes which their own desires initiate." He states further in another place that: "Their powers are excited into action by the necessities of the creatures which possess them and on which their existence depends." Again he states: "That from the first rudiment or primordium to the termination of their lives, all animals undergo perpetual transformations which are in part produced by their own exertions in consequence of their desires, aversions, pleasures, pains, irritations or associations." You will see then that all living structures are caused by reason of the desires of the parties who construct them, which desires arise from the wants and necessities of the parties. The desires will be likely limited to that party's experience or knowledge. Every action in life is traceable to an effort to adjust and adapt itself to meet conditions and external forces.

Professor Haeckel, the great German biologist states, "Cells are grouped together as builders or sculptors because they alone in reality build the organism." Still Mr. Haeckel claims that the actions of the cells are caused by merely chemical and mechanical forces,—that they are not intelligent beings. It seems absurd to me to claim that the living beings who are the sculptors and builders of all living structures have no intelligence. It is simply unthinkable.

In reference to the general course of development of animals he states: "For example—from the fact that the human egg is a simple cell, we may at once infer that there has been at a very remote time a unicellular ancestor of the human race, resembling the amoeba. Again from the fact that the human embryo originally consists merely of two simple germ layers we may at once safely infer that a very ancient ancestral form is represented by the two layered gastraea. A later embryo form of the human being points with equal certainty to a primitive worm like ancestral form, which is related to the sea-squirts or ascidians of the present day.

"If we go back to still earlier stages of development we are unable even to discover any distinction between the embryos of these higher vertebrates and those of the lower, such as the amphibia and fishes. Finally if we go still further back to the construction of the body from

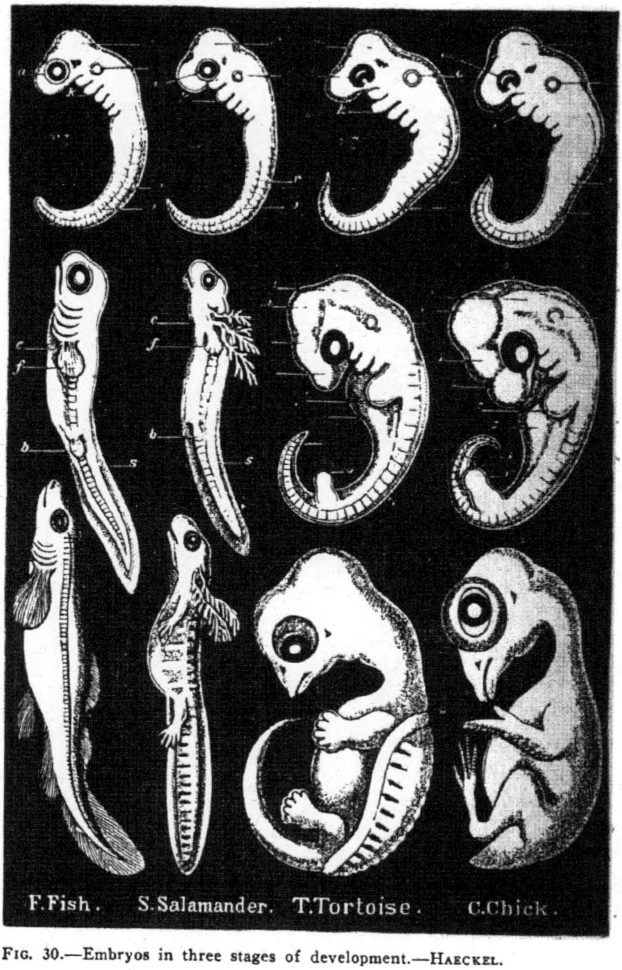

FIG. 30.—Embryos in three stages of development.—HAECKEL.

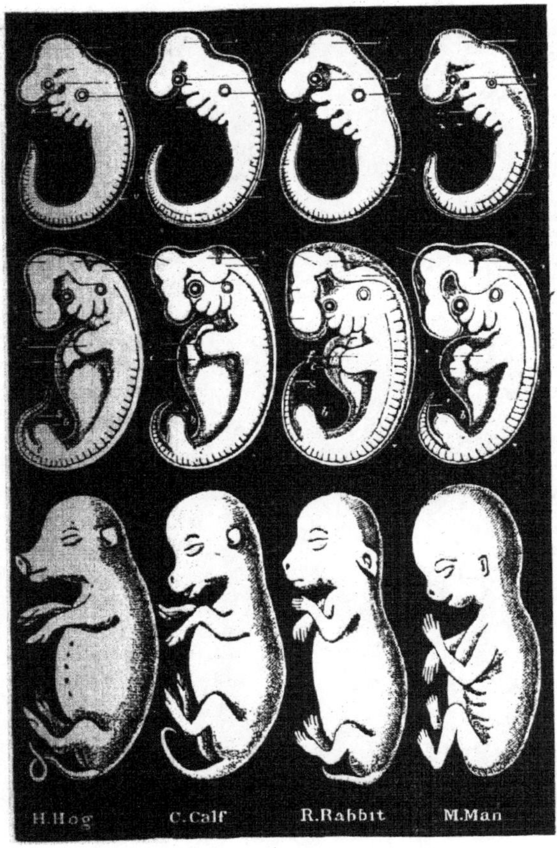

FIG. 31.—Embryos in three stages of development.—HAECKEL.

the four secondary germ layers, we may make the surprising discovery that these same four germ layers exist, not only in all vertebrates but also in all higher invertebrates and that they are everywhere concerned in the same way in forming the fundamental organs of the body. And if then we inquire into the origin of these four secondary germ layers, we find that they develope from the two primary germ layers, which are identical in all animals with the exception of the lowest division, the protista. Finally we see that the cells which compose the two primary germ layers universally originate by fission from a single simple cell, from the egg cell."

You will notice from this statement that the course of development or building of an animal, including man, takes place in the same manner in one animal as in another, and his illustrations clearly show the similarity of all animals in the lower stages of development. Figs. 30 and 31 show how the cell begins with the small and simple and gradually builds the complicated living structures. You see in the turtle how the cells have outlined the shape and frame of the shell. It shows that every move is for the purpose of building a moveable cell colony, which shall be covered with a protective armor. In the same manner you can see a purpose and intention of the builders in any other structure, as soon as it is partly completed, just as you can see in a partly completed house or automobile the intention of the builders. You can see a purpose in every act; every brick and piece of material must be placed exactly where it belongs; every part of the machine must be placed with a purpose and intention of working in harmony with every other part.

A being without intelligence could never work in this way with a purpose and with an intention of producing a work of art. Intelligence and skill of the highest order is required to produce these well balanced and highly complicated structures we call animals and plants. The common frog is a good illustration of the course and general development of an animal. He begins as a single cell which multiplies into enormous colonies which again group themselves into a fish. This fish gradually changes into the perfect animal or frog. This shows how the cells start building with a purpose and continue as far as

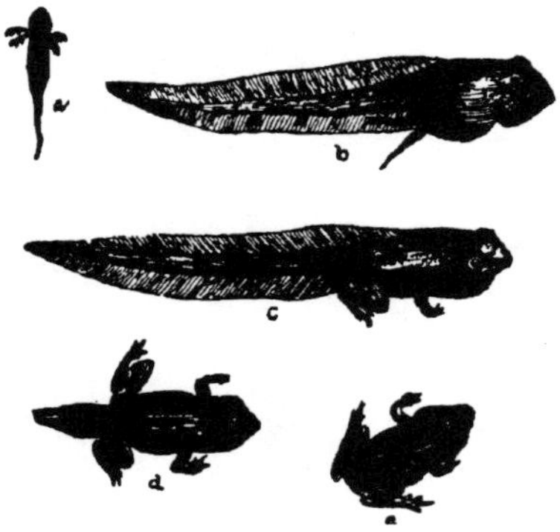

FIG. 32.—Tadpoles and Frog; *a*, tadpole with branching external gills; *b*, gills absorbed and hind legs have appeared; *c*, fore legs have appeared; *d*, tail shrunk and legs enlarged; *e*, perfect, young frog,—tail entirely disappeared. The figures represent some stages in the life history of the frog.—SCHUTE.

they have had experience. This particular kind of cell has been building these structures, and that is the extent of its skill and experience. In the same manner some people or animals have had experience in building certain kinds of structures and houses, and they always build those and none other because they are the only kind they know how to build. That point will be more fully considered under the cause of inheritance.

It must be clear to the reader by this time that all

structures in life are made for a purpose in a similar manner as all structures put up by man are built for a purpose. Living structures are made to move in air, earth or water or to remain stationary on the earth. In the same manner we build structures to move through the air, over the earth and water. It requires intelligence in the builder to produce these structures. For instance, to build a wagon, the builder must know how and where to obtain the wood, iron and other material, and he must have an idea of what he is about to do. Every move must be correct and for the purpose of completing the wagon. He must have knowledge of the strength of materials in order to produce a wagon that will carry the ordinary load. In other words to produce any structure designed for a purpose, requires intelligence. If it is to be a land animal, it must be built to occupy and move over the surface of the earth in a certain way. There are enemies, weather and elements to contend with, and these must all be considered in the making of the machine or moving habitation. The experience of the past must be there to guide the builder. From those experiences he must form ideas that guide his actions. The ear is made for hearing, just as a knife is made for cutting. These are both instruments made for a purpose and are produced by intelligence and not by chance. The eyes are made for seeing, just as opera glasses are made to help us see better. The cell completes the structure little by little for a purpose, just as if it had a model before it. What is the difference between the act by man of building huts and houses and the same act by the muskrat or the beaver. Consider how they all work in harmony for a purpose just the same as man. The birds build nests and a home to raise a family, so does man; the birds sing and display their beauty to attract the opposite sex, so does the human race. The bird is made to sail through the air. Consider how perfectly everything is calculated and adjusted to effect that purpose. The bones are all made hollow and extremely light, even lighter than aluminum. Those little animals we call cells build and produce these structures and attend to every detail in their production. Think of the factors entering into the construction of the eye. It is made to receive an image like a photographic plate with the purpose of transferring impressions and images received to the brain cells. Every step in the production of the eye must be exactly calculated, and every act performed with a purpose in view. Think of the perfect adjustment of the crystalline lens, vitreous humor, and cornea of the eye and how the light is continually regulated by those cells who have charge of the opening and closing of the pupil of the eye to adjust the intensity of the light falling on the retina. In reference to the development of the eye Mr. Haeckel has the following to say:

"The essential difference between the real eye and a part of the skin that is merely sensitive to light is that the eye can form a picture of objects of the outer world. This faculty of vision begins with the formation of a small convergent lens, a bi-convex refracting body at a certain spot on the surface. Dark pigment cells, which surround it absorb the light rays. From this first phylogenetic form of the organ of vision up to the elaborate human eye, there is a long scale of evolutionary stages—not less extensive and remarkable than the historical succession of artificial optical instruments from the simple lens to the complicated modern telescope or microscope. This great "wonder of life"—the long scale of the evolution of the eye—has an interesting bearing on many important questions of general physiology and phylogeny."

Just note how Mr. Haeckel compares the acts of the cell in perfecting the eye with the acts of man in perfecting and developing the telescope and microscope. The following statement also by Mr. Haeckel seems to me quite a conclusive admission by him of intelligence in the cell mind: "The history of civilization teaches us that its gradual evolution is bound up with three different processes:

(1) Association of individuals in a community.

(2) Division of labor (ergonomy) among the social elements and the consequent differentiation of structure (polymorphism).

(3) Centralization or integration of the unified whole, or rigid organization of the community. The same fundamental laws of sociality hold good for association throughout the entire organic world and also for the gradual evolution of the several organs out of the tissues and cell communities. The formation of human societies is directly connected with the gregariousness of the nearest related mammals. The herds of apes and Ungulata, the packs of wolves, the flocks of birds often controlled by a single leader, exhibit various stages of social formation as also the swarm of the higher articulates (insects, crustacea), especially communities of ants and termites, swarms of bees, etc. These organized communities of free individuals are distinguished from the stationary colonies of the lower animals chiefly by the circumstance that the social elements are not bodily connected but held together by the ideal link of common interest."

If the cell has gone through the same process of social organization and evolution as man, why is he not also the same intelligent being as man? Did you ever stop to think what takes place when the surface of the body is cut or bruised? The white blood cells, as they are called, who are the general caretakers of the body, whose duty it is to look after everything in general, such as the fighting of bacteria and disease germs and the general repair work, will sacrifice their own lives by thousands if necessary to save the body. They live in the body enjoying complete liberty. They do not float in the blood stream except when in a hurry to get somewhere, but move around everywhere as separate independent beings to see that everything goes right. If a bruise or cut happens they are at once informed in some way. We do not know just how, but they rush to the spot by thousands and direct the repair work and if necessary they change their own occupation and take a different job, that of making connective tissue in order to bind the tissues together. In nearly every open sore, bruise or cut, they are killed in great numbers in their faithful efforts to repair and close up the wound. A text book on physiology briefly speaks of it as follows:

"When the skin is injured the white blood cells form new tissue upon the surface, while the epithelium spreads over it from the edges, stopping the growth and completing the healing processes."

There seems to be no particular center in the body around which intelligence revolves. Every cell seems to be a center of intelligence and knows what his duties are wherever he is placed and wherever we find him. Every citizen of the cell republic is an intelligent patriotic being. It is a democracy where every individual enjoys an intelligent independent existence, and all are working together for the welfare of all. Nowhere can we find more absolute sacrifice of the lives of the individuals to the general welfare of all than we do in the cell republic. The results cannot be obtained in any other way nor at any less cost of individual sacrifice, so it is necessary to their social existence. The principle of individual sacrifice to common welfare has been accepted and agreed upon as the right thing and as their common duty, impartially distributed among them, and they perform their allotted work and duties regardless of their own individual comfort. I wish again to call your attention to the various intelligent acts of the cell, living as a separate individual before he has begun the social life. We find him then in the same place as savage man, before man began his social and civilized life. We find the cell using weapons like spears and bow and arrow with which to fight his enemy and capture his prey. I quote from Mr. Binet and others of the different actions of single cells as follows:

"With the cell the biologists can reconstruct the animal and vegetable kingdom by studying the forms and behavior of single cells and one celled animals. One can better understand the structure and physiology of the highest and most specialized forms, even that of man—for as Geddes has remarked: "The functions of the body are the result of the aggregate functions of its cell and are explained by variations or phases of the activities of them. Food is taken by the protozoa into the interior of the body, the digestible portion assimilated and the portion of no use to the organism afterwards rejected. The character of the nourishment also varies. Some forms live on vegetable productions alone, while others absorb any organic body, animal or plant, often devouring rotifers, worms or crustacea far higher in the scale than themselves. In the higher protozoa the food is either brought to the part of the body set aside for the reception of food by currents of water, created by the rapid moving cilia, while in others the animals which are eaten are in some unexplained manner benumbed by the protozoon and then devoured. The hunter infusoria are frequently armed with trichocysts. Trichocysts are urtical filaments, which serve the animalcula provided with them to disable or wound other micro organisms.

"A large number of infusoria, the paramecia, the ophryoglene, etc., use their trichocysts as organs of defense. With other species of which we shall speak more at length, the trychocysts are organs of offense. They are located either in the sides of the mouth or in parts adjacent thereto. This is the case with the lacrymaria, the didinium, the enchelys, the lagynus, the loxophyllum and amphileptus. These latter animalcula attack live prey that constitute their food in the following manner: They dash upon their victim and bury the trychocysts with which they are armed into its body. The victim is immediately brought to a halt whereupon the hunter siezes it and swallows it."

There are a great number of species of single cells which have invented weapons with which to fight their enemy at a distance. These cells, that make weapons and hunt their prey and also use them in defense, resemble man in his savage state very closely. Still they are microscopic beings, similar to the amoeba and those cells that build animals and plants. It seems to me absurd to say that these individuals, (whether as large as a mountain or smaller than a grain of sand), who display ability to invent, make and use weapons, who organize themselves into high class societies and republics, and build all the various living structures that we see are not endowed with intelligence. The following is a description of the cell who builds the human body from Conn and Budington's Advanced Physiology now used generally in the high schools and universities. "The amoeba is one of the simplest animals and lives in stagnant water. It is only a lump of jelly about 1,000th of an inch in diameter yet it is a complete animal for it moves, eats and grows and produces other amoeba." Man is like an amoeba. Each part of man's body is made of a multitude of living beings, each of which eats and grows like an amoeba. Each tiny being is called a cell. One collection of cells form the skin, another the muscles of the arm and another the stomach and so on through the body. Each collection does its own work without interfering with the others. The cells work together like a well trained army, so we do not feel the workings of each separate cell. If a collection is out of order the person is sick.

"Relation of cells—In the body formed by the cells there exists a controlling spirit of life, whose nature is unknown. When all the cells are obedient to its influence, the body as a whole is alive, but if the cells are not obedient, the body as a whole is dead, although each separate cell may remain alive. For example, a blow upon the head may disturb this controlling influence so that it cannot tell the cells how to act. Then they instantly stop work and the body drops dead. Yet each cell may remain alive for minutes or hours just as each soldier may remain alive after an army has been disbanded." This description of the cell colony or individual we call man is very good, as it illustrates its high state of organization.

Before I close this chapter on structure, I must call attention to one of the wonderful fly catching plants that grows in the bogs and swamps of South Carolina. It is called Venus Fly Trap. This plant, which has roots and obtains nourishment from the ground and air like other plants, can manufacture its own food from the earth, air and water like other plants by the aid of sunlight and does not need flies and insects for food. It catches insects as a mere pastime, sport and luxury. The fly trap is the most scientifically built structure that could be conceived to effect the purpose for which it is made. It opens and closes just like the jaws of a steel trap. The trap is built out at the end of the leaf showing that it was a later idea conceived by the builders of the plant after having first lived for ages without this trap. The idea of catching a fly now and then for food or sport or amusement must have arisen in the minds of the builders of this plant. The complete fly trap must have been first conceived by the builders before they could have entered upon the work of

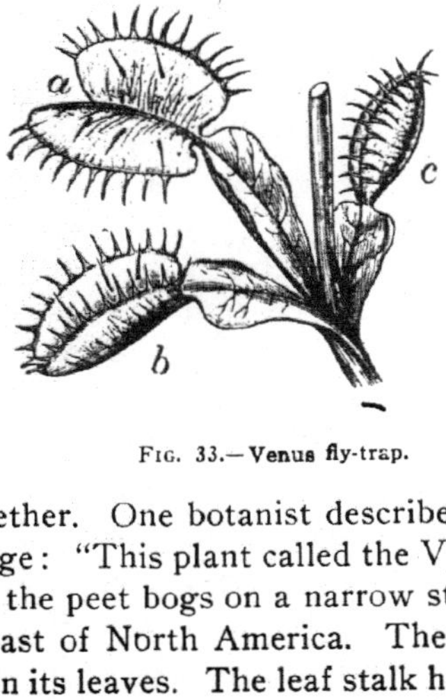

FIG. 33.—Venus fly-trap.

putting it together. One botanist describes it in the following language: "This plant called the Venus Fly Trap grows only in the peet bogs on a narrow strip of country on the east coast of North America. The peculiarity of the plant lies in its leaves. The leaf stalk has become flattened out so as to be leaf like, while the blade proper has become edged with teeth and has moreover six sharp little bristles standing straight up three on each side of the midrib. These midribs are sense organs. Touch one ever so lightly and the halves of the leaves on which they are placed close up together abruptly just like the slamming to of a volume, the midrib serving as hinge while the teeth at the edges interlock like clamped fingers. The sense bristles too shut up, as a blade of a pen knife closes. If the touch which evokes this response has been given by say the end of a pencil the two halves of the leaves will slowly open again and the bristles raise themselves but if some insect brushes against them; then the rapid closing of the leaves makes it a prison, while out of the glands of the surface of the leaves a digestible fluid quickly overwhelms the poor victim. When the nutritive parts are completely absorbed, the six sense bristles once more stand erect ready for action like soldiers on guard. Indeed it is an open question whether in the whole animal world there is a more perfectly constituted organ of touch than is found in the dionia or Venus Fly Trap."

I have myself examined a number of these plants and the general description above is correct. On each side of the leaf are three sentinels on guard,—simply hairs or feelers to give notice to the thousand or more individuals to act as one in slapping the trap shut and catch the insect when he gets inside of the dead line. Consider the situation here—a colony of cells in the shape of a plant. Here is no brain, nerves or muscles, as we understand them, simply a plant. Still it is a colony of cells in the same manner that an animal is a colony of cells. Insects are not absolutely necessary as food for the builders of this plant—still they have devised and built an elaborate and effective machine with which to catch flies. Considering the size of the fly trap enormous numbers are required to perform the work of slamming the trap shut when notified to do so. They are not fooled by anything, and the trap doors only shut upon insects or food. It is inconceivable that the builders and operators of this fly catching machine can be anything but conscious and intelligent beings. Where does the chance theory of natural selection and survival of the fittest of the evolutionist come in here? That theory claims that everything came to be as it is by accidental variations beneficial to the being, that is to say, every beneficial variation would make him the fittest in the struggle for existence, and he would by reason of such beneficial variation survive and perpetuate his kind. However, in this case, as in all of the cases I have examined, the variations would not be of any benefit until the structure towards which the variations tended was completed and in working order. No chance variation could ever have produced this fly trap. Any variations or accidental changes in the leaves of the plant could never have produced the fly trap and could not have been a beneficial variation until the fly trap was completed, but on the other hand, rather an injury. It is true, of course, that in the struggle for existence between individuals, the best one will win, but that does not prove or in any way show how those individuals came to be. For instance—two battleships meet out at sea in mortal combat, the best ship will likely win out and will be left to produce more ships of a like kind but it does not show in any way who produced the ships. What I want to get clear is this, that the survival of the fittest theory is only an incident that determines as between two individuals who shall live, as there is not room for both, but it does not show the cause of the existence of these individuals. With the microscope that we now have, we can see what the cause is, and who are the builders of all those living things we call plants and animals, and it seems to me that it is time to get down to facts and admit the truth as far as we can see it. What process of development took place in the past ages to produce the cell we have today, is another question. How the primordial cell of which the cell is composed came to be and multiplies is another question which we are not at this time able to answer. It was just as much of a mystery 35 years ago what produced the plants and animals as it is now a mystery what produces the cell. I believe that some day the mystery will be solved.

CHAPTER V.

WHAT IS INTELLIGENCE?

What is understood by the word Intelligence? What should be considered an intelligent being or individual? This must be settled before we go any further.

First, what is understood by intellect? The dictionaries define intellect as, "That faculty of the human mind which perceives, understands and thinks." Webster states that it is the "Faculty of the human soul or mind which receives or comprehends the ideas communicated to it by the senses."

These definitions clearly show that the sense organs, eye, ear, nose, etc., are no part of the intellect. The intellect then is the brain cells. They receive impressions and think.

How do the dictionaries define intelligence? They define it as, "The power of discovering or understanding." This definition includes both the senses and the brain.

To be able to, and to have the power to discover you must not only have the brain cells to do the thinking, but you must have the apparatus to receive and transmit to the brain cells, impressions and conditions existing in the world outside of the body, so they can have something to think about, or something to understand.

The dictionaries define intelligent as well-informed and skilled. A person could not be well-informed, skilled, or intelligent, unless he possessed most of his sense organs and also his brain cells or thinkers.

The five senses are the instruments by which the brain cells obtain this information from the outside world. They have nothing to do with the thinking.

The brain cells and their assistants, the nerves, have charge of the business of receiving information from, and being in touch with the outside world, and of directing and guiding all the other members and cells of the body. This will clearly show that the cells may be ever so able to think and direct,—that is intelligent, as far as they are concerned, but should they be deprived of information from the outside world, they would not be able to exercise their faculty of thinking and directing, especially in reference to matters taking place in the outside world, be they ever so dangerous to the existence and welfare of themselves and the whole body in general. To illustrate: If we left the house and went into the timber with a gun to shoot a lion, and the gun should accidentally explode and destroy the eyesight, we would likely have a hard time finding the way back home, and would probably be eaten by the lion. Why? Because the instrument by which we convey continually to the brain cells information of the outside world, so that the brain cells can think and know what is going on outside, has been destroyed, and they have no means of knowing of any impending danger, or of where to go and cannot give any directions to the motor or any other part of the cell body.

It is the business of the brain cells to inform the other cells of the body of the approaching lion, and where to go, and to order the limbs to immediate action, in order to save the whole body or cell community from utter destruction by the lion. The brain cells are, however, helpless, because the instrument necessary to receive and transmit the information is not working. If a person at a distance saw a man standing on the road with a gun, and a lion coming toward him, and this man made no effort to shoot or to get away, the person would declare that the man was crazy and not possessed of any intelligence; and yet the man, or the cells of his brain, would be just as intelligent as ever.

To further illustrate, if a submarine should start out from Germany and by accident should destroy its headlights and periscope, it would not be likely that it would get back home without being destroyed by an English cruiser or battleship. When it came to the surface for air, like the whale, it could not tell whether it was near an English destroyer or not, nor could it see the dangers ahead, when starting towards the steel nets stretched out ahead of it.

There is no difference whatever in the purpose and functions of the periscope of the submarine and the eye of a man, animal, or fish. The man at the other end of the periscope gets a picture of the situation in the outside world and from this picture he thinks, reasons and decides what to do. If the battleship is too near, he orders the different acts to be done by the parties in charge of the propellers and rudders, in order to escape the enemy. In the same way, in man the brain cells at the other end of the eye get a picture of the situation and if they observe an enemy approaching, or too near, they order immediate action so as to get out of danger. The same is true of a fish,—like a submarine, it would order the cells in charge of the propelling apparatus and rudder to action to move to a place of safety and away from the enemy.

This should clearly illustrate the proposition that the senses are instruments which are constructed for a special purpose of obtaining and transmitting information from the outside world, and have nothing to do with the thinking or reasoning, which will be necessary to intelligently direct the actions of the other organs of the body to move to a place of safety.

We see them in every organized body or being, like animals (and we shall see later also in plants), where there has been a division of labor and a special organ arranged for receiving and transmitting sensations or information, that the bunch of cells which have any special work in charge, like the cells which make up the eye and ear. have nothing to do but receive and transmit impressions.

The thinking and directing is done by another bunch of cells which we call the brain. The two things are necessary in an intelligent animal, the brain and the senses, that is, the thinkers or the brain cells, and the senses or the cells engaged in, or in charge of the instruments for receiving and transmitting information regarding conditions existing in the world outside of the body.

There must always be the thinkers and the transmitters of information to the thinkers. We shall see, however, later, that the individuals who have charge of the receiving and transmitting of information are and must necessarily be intelligent beings as well as thinkers. It requires intelligence to execute orders received as well as to give the orders.

It is a peculiar trait of the human being to consider every one who does not agree with him in his ideas of any civil, political or religious belief, foolish, ignorant and not intelligent. We look upon the savage as being inferior to us in mental capacity, when in fact investigations lately have disclosed the fact that he is just as intelligent in his occupation as any civilized man. I quote from the Scientific Journal the following:

"Civilized man has so long scorned the savage races as

mentally inferior, that it will be hard for him to become accustomed to the idea that even the repulsive Fiji Islander is his equal in intelligence.

Yet this is exactly what is being found to be a scientific fact, and the world must accept it if we are to have any success in civilizing the lower races.

According to this revolutionary idea the difference between us and the savages is not one of mental capacity, but rather of the objects upon which that capacity is exerted.

A man may, for example, display just as much intelligence in tracking a kangaroo through the bush as in solving a problem in higher mathematics.

What makes the savage a savage is not the lack of intelligence, but the fact that his intelligence is exerted largely upon foolish superstitions. The contents of his brain differs from ours in kind, rather than breadth and depth.

"Dr. Alfred G. Mayer, of the Carnegie Institution, found some interesting evidence to support these arguments among the Fiji Islanders, who are the lowest, most cruel and most repulsive of primitive savages.

I remember how my parents used to consider the Catholics a very ignorant and uneducated lot of people, and the free thinkers, as we called them, the most ignorant of all, and also dangerous to associate with.

As I came to maturity, I soon noticed that it was only a matter of opinion and that one was about as smart as the other.

My father, who read only republican papers, was a republican, while my brother and I happened to read some democratic literature and we turned out to be democrats for a while, but later saw that it was all a matter of politics. Whether any person is really sane, is a question; sometimes it looks to me as if we all were walking along the borderland of insanity, hysteria, dementia, idiocy and imbecility, for most of us seem to be afflicted with a mental weakness of some kind.

I believe that the cause of mental weakness in man arises from over development of the mental faculty, the brain. It becomes too complicated and the wires get crossed. Consider all the religious beliefs; the foolish patent medicine schemes, faith cures, get rich quick, spiritualist and other superstitious nonsense, which rope in people. Consider how the asylums are being filled with the insane of all kinds, people morally insane, intellectually insane, as monomania, people afflicted with common mania or raving madness, dementia, imbecility and idiocy. Very few human beings are capable of reasoning, as we understand reason.

Think of the superstitions of the past centuries, the burning of the witches, etc. The following article from a newspaper is significant of the prevailing condition of the human mental machinery: "The next time you are inclined to scoff at any person for indulging in 'superstitious' beliefs and practices, just stop a minute and consider whether you are not in some degree superstitious yourself. If you make an honest self-examination I have not the slightest doubt that you will be more charitably disposed toward the 'poor fool' whose belief in ghosts, witchcraft or charms you were about to ridicule.

"Not long ago there was a notable trial in Boston of a man accused of selling 'lucky stones.' In the course of his trial, which ended in a conviction and jail sentence, it developed that for several years he had done an enormous business in these magic gems. His patrons, who annually numbered up in the thousands, were by no means confined to the ignorant and the uneducated. It was found that men and women of prominence, people of real intellectual power, had contributed to swell his profits, and some of these people had even sent him letters testifying to the 'good luck' which his 'lucky stones' had brought them. Personal observation and research have indeed satisfied me that education is as yet far from being an absolute antidote for superstition.

"I once closely questioned a number of Harvard university professors and instructors to ascertain their freedom from, or indulgence in, superstitious customs. I found that, with only two exceptions, every one of the highly educated men whom I questioned confessed to having some pet superstition. Several had the habit of 'knocking on wood' to avert misfortune after a boastful remark. Others could not pass a pin without picking it up. Others believed it bad luck to go under a ladder. One eminent historian, seemingly as matter-of-fact and hard-headed a man as could be met anywhere, confided to me that he felt quite uneasy unless he was standing in a certain position when he caught his first glimpse of the new moon. Other investigators—notably Prof. Dressler, who conducted an extensive census of superstitions in California—have found the same state of affairs. It would seem that, no matter how well educated a man may be, he usually has in him a 'superstitious streak.' "

This indicates prevailing conditions, how the brain cells of the human being are limited in their capacity to comprehend matters rightly. There is no question about it in my mind, that the cause of the mental weakness in man is the abnormal development of the brain. We find very little, if any, insanity among animals. They pursue their line of work with the keenest intelligence as far as pertains to their business in life.

I did considerable trapping and hunting in my younger

days, and have watched the musk rats construct their houses for the winter in the marshes; and it is a wonderful work, when you take all the conditions into consideration. The following is an article which I cut from a magazine today concerning the beaver:

"That the beaver is a capable engineer from his own point of view, and an indefatigable worker to boot, has been impressed upon us from our infancy and is enshrined in our daily speech. Every one knows what 'working like a beaver' means. But few of us have realized that the beaver's engineering feats are of benefit to us as well as to himself. The beaver 'works for the nation,' as Harvey Ferguson put it in an article contributed to The Technical World Magazine (Chicago, July). 'He is one of the most useful irrigation engineers we have,' for he builds dams; not very big ones, but a great many small dams will impound as much water as one large one. His work is now considered so valuable, Mr. Ferguson tells us, that he is to receive special government protection. We read: "The beaver is solving one of the problems which has proved most vexing to engineers since irrigation became prominent in the West. That is the problem of water-supply; for while engineers can construct dams and ditches, they usually must rely upon nature to furnish the water for the project. And that is where the beaver helps —he sees to it that the supply of water is maintained.

"The beaver does not intend, perhaps, to do all this; he is simply interested in securing a lake in which to live. But while he serves his own ends, up in the mountains, he also impounds enough water to insure a constant supply for irrigation-projects, and that is why he has won governmental favor. Officers of the Forest Service, who are doing all in their power to protect the beavers, and scientists of the Geological Survey, who have studied the work of the animals, say there is no doubt but that these beaver lakes will be of great practical value."

The brain cells of the animals are not so numerous as they are in the human being. They are more simple and not likely to get mixed up with each other's business.

We pass through life without taking any notice of the keen intelligence displayed by the birds and animals, as far as it pertains to their life. You all remember how the prairie plover and the prairie chicken and partridge will act when you approach them on their nest. They will first sneak off to one side of the nest as far as possible, then pretend they are wounded, as if they had a broken leg or wing. It fooled me at first to such an extent that I did not find their nest when I went back to look for it, as I followed them too far before I got wise to their trick. I noticed that my dog did not get wise, as he would chase them every time as far as he could. The manner in which they will simulate a broken wing or leg is certainly perfect. Sometimes one wing will hang down loose to one side just as if it were dead. You never hear any one mention this common trick practiced by birds to fool other animals. A trick is described in the following article which I read the other day, which is identically the same trick as practiced by the birds:

"Two boys, one the possessor of a permit, were fishing on a certain estate when a gamekeeper suddenly darted from a thicket. The lad with the permit uttered a cry of fright, dropped his rod and ran off at top speed. The gamekeeper was led a swift chase. Then, worn out, the boy halted. The man seized him by the arm and said between pants:

"Have you a permit to fish on this estate?"

"Yes, to be sure," said the boy quietly.

"You have? Then show it to me."

The boy drew the permit from his pocket. The man examined it and frowned in perplexity and anger.

"Why did you run when you had this permit?" he asked.

"To let the other boy get away," was the reply. "He didn't have none."

The human mind like the animal mind is the united action of the brain cells. In the human brain they number several hundred million. It is the impression and information received by these cells and stored away for future reference which make up the human mind. This power and ability of the brain cells and all other cells to store away information for future use we call memory, and upon this power all reason and judgment is based. Without memory there could be no judgment or reason, as we understand these words. We shall see later that all the cells of the body have the power of memory.

It is now being recognized that many lower animals and birds have keener senses than man, such as smell, hearing and sight. I can remember from experience in hunting geese from pits, that their sense of sight was considerably better than mine. The following also seems to indicate that the sense of hearing is keener in birds than in man:

"British journals have commented on the strange disturbance among the pheasants in northeastern England while the naval battle in the North Sea was in progress on January 24. The sound of the cannon could not be heard in many places along the English coast, but it is evident that the birds heard it and were greatly agitated by it. This is another interesting proof of the fact that keen as are the senses of men, those of many of the lower animals are much keener.

"The human ear can not hear sounds that are produced

by sound-waves faster than about 20,000 to 28,000 to the second. The ant, as has been demonstrated, can hear sounds produced by a higher vibration, and for aught we know animals of higher organism than insects have as keen ears. They have a great advantage over us. And if they have sharper ears they also have better eyes and noses."

Now, whether or not their judgment or reasoning ability is any better, will be hard to tell. The sense organs are the instruments constructed by the cells for specific purposes, each for a specific purpose, the eye to take and convey to the brain cells pictures of the surroundings; the ear to take vibrations from the air and convey them in the same manner; the nose to catch particles of matter floating in the air, analyze them chemically, and transmit results of the analysis to the brain cells. It would appear that the cells of one animal are about of the same intellectual capacity as those of another. The reason that the cells of the dog are able to produce a better smelling apparatus is because they make smelling a business specialty. The effort is concentrated upon that point and his eye and ear are very inferior to that of the bird. Again you would not say that the bird is more intelligent than the dog, because it has better eyes and ears, nor would you say that a dog or a bird is more intelligent than a human being, even if it has better eyes, ears and nose. Still you would have to admit that dogs and birds act with as much intelligence in their place in life as the average man.

We now come to another intelligence called instinct, impulse and emotion. Most of the writers on psychology do not allow for animals any power to reason, but claim that animals are simply automatons, that they act from what they call instinct. They claim that the acts of all animals are instinctive, emotional and impulsive and are not caused by intelligence and reason.

As man is an animal and performs as many instinctive acts as any other animal, we can easily clear up some of the mysteries.

We will take the mating instinct, to begin with. Every one is well acquainted with the male desire for the exclusive possession of a certain female. This desire comes on at about seventeen or eighteen years of age. Before that time he had not the slightest desire for the possession of, nor even the company of females, he preferred rather, the company of the male sex. At this age also peculiar masculine features appear, such as the beard, extra masculine developments and change of voice. This change also takes place in other animals. From whence comes this desire?

Nothing in the world seems more perfect in form, sweet in disposition, and beautiful of face than this particular female, to which he happens to take a fancy. Heaven on earth seems to be in sight, by obtaining possession of her.

Now who and where are the parties who seem to be continually arguing this proposition to him? He seems to believe and feels fully convinced of these facts, so much so that it is almost impossible to demonstrate or prove to him anything to the contrary. We know now that the beings who are responsible for these opinions and actions are the cells who have the matter in charge of building a new human being. These cells of the body who have in charge the work of perpetuating themselves and the human race, which is after all the most important work of all, are called sex organs. We know now if these are removed, no desire for the female will arise.

What method do these cells of the sex organs employ to make him(the brain cells) feel and believe that to marry a certain female is the only right and proper thing to do? These cells in control of the perpetuation of the race take possession of the entire body and compel it to act in a certain way. They get control of the mind, that is of the brain cells, and in that way get control of the entire body. These actions of the sex cells or nerve centers of the body, directing our actions, are called the mating instinct or instinctive acts. There are a great number of these nerve centers in the body, which give the body a large number of instincts or instinctive actions. It is the same in man as it is in animals. They all have their instinctive acts. These acts are performed with the will of the brain cells, but the will does not originate in those cells.

At the very beginning of the construction of the human individual, these cells are set apart and placed in charge of the matter of the building of a new individual and the perpetuation of the race. These cells are to attend to nothing else. As an illustration take a submarine boat. The submarine is an individual. We determine from the actions of the submarine whether it is or is not an intelligent individual or being. The submarine can only act according to the will of the people in charge. The same is true of man or an animal. It can only act as directed by the beings in charge, which we call cells. The builders of the submarine, if they should want the existence of it perpetuated, would do the same thing as the builder of the human being. They would leave the whole matter to a special committee to take charge of that entire matter, and to look after that and nothing else. This we find to be exactly what takes place in man. The intelligence of the boat or submarine is the combined intelligence of all the individuals of which it is comprised and which have it in charge. The same with the human being, animal or plant. The intelligence of the plant or animal will be the combined acts or intelligence of the entire community, which we call the individual or animal.

Now we shall take up some of the different and special work taken care of by special committees or nerve centers who have been assigned to look after it, such as breathing, circulation, digestion, etc. Prof. Harris, after describing the brain, makes this statement:

"Here are situated many of the most important nerve-centers, or collections of nerve-cells, superintending such bodily activities as breathing, the heart's action, the regulation of the size of the blood-vessels, perspiration, the flow of saliva, the flow of gastric juice, the chewing of food, swallowing, voice-production, and the act of vomiting; and there are other centers still.

"In a later chapter we shall discover what a center is and what it does. Higher up the brain stem are centers for facial expression, eye movement, iris (pupil) movements, focussing of near objects and the secretion of tears. In the brain proper cerebral cortex are centres or areas underlying such states of consciousness as sensation, perception, memory, and emotion, besides speech and voluntary movement."

This is a plain statement of the facts, how collections of cells have been placed in charge here and there all through the body to superintend this and that particular work. These cells or collections of nerve cells are called little brain or nerve centres. We have the same conditions of affairs in other large organizations of individuals, such as a battleship, German army or telegraph system.

The spinal cord is comprised of a great number of such nerve centres. Prof. Harris states: "The spinal cord is, however, something more than a conductor or transmitter of impulses; it is also a collection of centres in series from above downwards. Nerves enter the cord and nerves leave it at certain levels all the way down. The ingoing nerve and the outgoing nerve are anatomically and functionally linked in the interior of the cord; this place is called centre. A centre on its structural side consists of at least one cell on or over which the ending of a nerve fibre is distributed; usually, of course, many cells go to make up a single centre."

Then he makes the following statement: "But what, in a few words, is the nervous system for? It is first of all for carrying out certain activities, such as breathing, on a sub-conscious plane, over which the constant supervision of consciousness would be tedious, if it were not impossible, considering the enormous number of demands made on the attention of the individual. It is in the next place for carrying out certain activities with the greatest possible speed compatible with the greatest possible accuracy; it is for linking up functionally the outer world with our living bodies, allowing it to act on our bodies within certain limits, and reversely acting itself on the outer world, when and to what extent may be found necessary. It is for maintaining our posture in stable equilibrium, whether we are sitting, standing or walking, much of this it also does on the sub-conscious plane, the cerebellum being the great central organ for attending to this important but not necessarily always conscious affair.

"The element of speed is a very obvious one in the activities of the nerves; it is of supreme moment that you drop the red-hot coal as soon as you possibly can, that you wink away the grit in the eye almost the instant it alights. It is of great importance to the artilleryman to fire off his gun the instant he gets the order from the officer to do so, a fraction of a second's delay may mean that he misses the target. It is because the nervous system of the active young man works promptly that he skips out of the way of the motor car, whereas the old gentleman with his much more slowly acting one may be run over. If you are too slow at taking the 'service' at tennis, you will miss the ball. But the nervous system attends to other matters than mere speed; it cares for accuracy too. Thus at tennis not only must I strike the ball at the right instant, but at the proper spot of my racket, not the handle, not the wood. The muscular adjustments necessary to bring the center of the racket in contact with the flying ball are carried out by co-ordination. A large number of muscular activities have to be co-ordinated to bring about any desired activity; many, perhaps widely separated, muscular groups have to be brought into simultaneous contraction, while other muscular groups require to have their activities restrained. All games of skill, for instance, billiards, gymnastics up to trapeze displays, and tight-rope walking, involve this co-ordination. The nervous system is the essential in co-ordination, whether it be in the baker's boy balancing his bread-board on his head, or in Cinquevalli balancing two billiard balls on the top of a cue on the end of his nose. Supreme accuracy of muscular adjustment it is that underlies all these different performances, but in smaller degrees we all employ it, as we are bound to do. We co-ordinate our muscles to step out of a carriage on to the ground, to ride on horseback, to ride a bicycle, to steer a motor car, to use a typewriting machine, and so on.

"Besides accuracy of adjustment there is the element of the precise amount of force to put into the muscles. This putting forth the exact amount of force—neither too much or too little, means the exact or right measure (of force). Thus not only must the tennis ball be struck at the right instant with the right spot of the racket, but it must be struck with the proper amount of force in order to be returned 'within court.'

"The nervous system as a rule learns by slow and painful experience. We must use the exact amount of force in order to accomplish any result with accuracy. In ordinary walking we learn to put forth exactly the necessary amount of energy, neither stamping heavily on the pavement nor putting down our feet so feebly that we would make no progress at all. * * *

"But it is very clear that all this precision, co-ordination, and expenditure of the correct degree of muscular energy is only possible in proportion as the muscular and other senses are properly trained. * * *

"We see, then, that the nervous system puts us into communication with the outer world and its inhabitants which act on us, enables us with speed, accuracy and the correct amount of force to react upon it, and then it makes us aware of our own bodily position to the changing states of the environment. In our nervous system we store memories of what has happened, we register experience for the future, we communicate as we will with our fellow beings and, maintaining our self-conscious identity, we continue our conscious connection with the past. Nerves and the nervous system not only protect the individual from injury, enabling him to seek food, avoid or overcome enemies; but they are constantly handing over some activity or other from the conscious to the subconscious realm. We educate the nervous system laboriously to perform certain actions, conscious attention being very much concerned in it. The acquisitions are relegated to the unconscious or at least subconscious realm and are at last carried on without the interposition of attention at all. There is a very great saving of nerve energy here; things so done are called habits. Such coordinated activities as the maintenance of posture in walking are, in this way, carried on below the conscious level, so that as we walk alone we can be engaged in solving a problem, or if with a friend we can carry on conversation without having to give any attention to the movements of the limbs. Even talking can become an automatic affair of this kind, as when we recite a poem without thinking of each word, and what comes after it, as we had to do when we learnt it originally. Habit is the popular word for all these activities which may or may not have been originally learned, but which are now all relegated to the subconscious sphere. 'Instinct' is the popular term for habit as found congenitally present. We say that a child knows how to suck by instinct; certainly it is not by consciousness, for children without brains at all (acephalic monsters) can suck perfectly. The child inherits the capability of carrying out the co-ordinated movements of sucking; it does not require to learn these, they are potentially present in its nervous system. We shall later see that this sort of thing is only an example of a certain kind of reflex action.

"One of the best illustrations of how the nerves work is the one so often given, the telephone exchange. What is a telephone exchange for? To put two people into (verbal) communication with each other."

This statement will illustrate what we mean by a nerve centre. It is as Prof. Harris states, a collection of cells, set apart for superintending, looking after or having to do with a particular function or activity of the body. We have collections of cells looking after breathing, if we destroy them the breathing stops. To illustrate,—if the continual pumping of air into the submarine was necessary to the lives of its inhabitants, the same as it is in the animal, a number of individuals would be placed in charge of that work, and if those individuals were killed, the pumping of air into the submarine would stop and the submarine would be killed—all life in the submarine would be destroyed for want of air. In the same way want of air destroys all life in the animal.

Our consciousness is the cells or individuals connected with the senses of sight, hearing and smell. We can see that in the submarine as in the human being. It should not be necessary that the cells of those special organs should be bothered with this special work of breathing or the pumping of air into the submarine; those in charge should know that to stop or neglect the work will mean the utter destruction of themselves and the entire individual.

Understand, now, that the nerve cells are not the cells that do the work, the muscles are the workers or laborers. The nerve cells are those in command. They are to keep the workers informed at all times, not only when to pump air, but how best to pump. Now you can clearly see that it requires just as much conscious intelligence to look after this pumping work as any other work. There must be a conscious intelligence in charge of any special important undertaking like this. Every collection of cells we call a nerve centre or little brain must be a sort of conscious intelligence. It must be conscious of its work in hand, in the same manner as we are.

The following statement by Prof. Harris illustrates how information from the world outside of the body is taken and transmitted to the cells within occupied with nothing but thinking, such as the brain cells, and how they may handle the information, and send out orders to certain special nerve centres, whose business it may be to take charge of this particular matter in hand. He says:

"Centres are, of course, not only executive; we have centres for the reception of impulses, whether these arouse consciousness or not. Thus we have centres for seeing, hearing, touching, smelling, tasting, and so on.

These sensory centres are highly specialized portions of the brain in which impulses from the end-organs of sense are received and are usually worked up in consciousness into the perception of an object, a sound, a thing in contact with the skin, a smell or a taste. These centres are specialized to receive, as the others are specialized to emit. But it is quite clear that no centre can emit and never receive. A sensory centre, in the first instance, receives and may for a time retain, but sooner or later it transmits impulses either to an executive centre or to another sensory centre. Thus when I see an apple, impulses not only pass to the centre for vision, but onward to the centre for taste, and from both these, impulses can go over to the motor cells in the middle part of the cerebrum, whence volitional impulses descend to the muscles of the hand prepared to seize the fruit. That this is the physical basis for the association of ideas there is little doubt."

To illustrate with a submarine, the end of the periscope of the submarine, which is the same as the eye of an animal, discovers a British ship in the distance, and gives orders which will cause actions to take place, like firing up the boilers, putting on more speed, twisting of the rudders, etc. One activity will lead to another. Now we shall give illustrations of the actions in case of some local trouble or disturbance. Prof. Harris states: "We get some grit into our eyes, and in consequence there is a great outpouring of tears. We are not weeping, there is no emotion calling forth tears, idle or otherwise, and we have certainly not willed tears to flow. The lachrymal glands have been reflexly stimulated to secrete. Clearly there must be accessible to incoming stimulation some specialized portion of the central nervous system, which is set apart for inducing secretion in the tear glands, just as we saw there was a similar region for the salivary glands. Such a specialized portion of the grey matter of the central nervous system is a center, in this case."

Now it is plain that if some gravel or other substance should get over the end of the periscope so as to interfere with the entrance of the light or getting a good clear view of the outside surroundings, it would be the business of the party in charge of that end of the periscope to remove the obstruction. Although it would annoy and irritate the parties at the other end, still it would be none of their business to remove it, in fact they are so far removed from the place that they could not attend to it if they would.

So the parties in charge of the other end of the periscope take care of the trouble, and remove the grit or obstruction without any orders from any one higher up. He procures water and washes it off in the same manner as the cells cause water to flow over the eye, with the purpose of washing away the grit. Any one can clearly see that every nerve centre, whether it is composed of one cell or a thousand, when it has charge of any special matter like this, must be possessed of intelligence. The only purpose of these nerve centres or "little brains," as they are sometimes called, is to receive information and give orders to other cells who work under them. No beings can take charge of the activities of other beings and guide their actions to do or not to do the right thing at the right time, unless they are intelligent beings. It requires just as much intelligence to capture a fortification as to give the order to do it. To sum up the nerve centre question, I shall again quote Prof. Harris as to what it is, and how each collection of cells attends to its own business and nothing else. He says:

"A centre is a group of nerve-cells so constituted that

when these cells are stimulated in any way whatever, only one kind of activity, some special activity, is the result.

A discharging centre is for an action, it presides over it; when stimulated, that action occurs more intensely, when inhibited that action is restrained or suppressed. When the centre is destroyed, the action is rendered impossible. The doctrine of centres affirms that there is specialization throughout the nervous system, that certain cells and these alone are concerned with the performance of a particular function, that if that group be destroyed some particular activity becomes impossible. If this functional specialization did not exist, then any one cell-group in the central nervous system could act vicariously for another, but this does not happen. The respiratory centre cannot act for the sweating centre, nor either for the vomiting centre. Just as a gland is not a muscle and cannot do a muscle's work, so the centre for salivation is not that for the flow of tears and cannot do its work. There is, in fact, higher specialization in the nervous system than elsewhere. One neural region governs the muscles of breathing, another the diameter of arteries, another the glands of perspiration, another those of gastric juice, another the act of vomiting, while still others are the places of the uprising of emotions, volitions, speech or memory.

"The central nervous system is not a neural chaos in which the units are unrelated or equivalent existences, but it is a cosmos in which the functional units are differently endowed and are related to each other after the manner of a hierarchy, the centres being arranged on several functional planes. Centres are co-ordinated by being subordinated some to others. There is no equality of functional position as there is no equality of capabilities. The nervous system knows no such thing as socialism, if by that is meant equality of position and powers. But in the nervous system there is a neural society in which there are aristocrats who rule and give orders, and servants who serve and obey. Higher centres control but do not domineer over lower. The doctrine of a neural hierarchy is one of the most luminous in modern neurology."

It is a wonderful thing how the nerve centres take care of the different special work here and there in the body, many times against our will—that is against the cells of the brain connected with our senses. Such acts as coughing or sneezing in church we would like to stop, but the tickling sensation in the nose or throat brings on the sneezing or coughing, in spite of our wishes. If it were not so, we would soon get in more serious trouble, when dangerous germs attack the membrane of the throat or nose.

The nerve cells who have charge of the work of preventing dangerous germs entering or getting lodged in the lungs or throat, and in that way finally getting into the blood stream and probably destroying the whole body, are continually on the lookout for these dangerous enemies, and when they lodge on the membrane of the throat or nose, in the act of breathing, they attempt to expel them or dislodge them by first secreting and enveloping them in a sticky fluid and then blowing them out entangled in the fluid, by the act of sneezing or coughing. There are nerve centres who have charge of secreting the sticky fluid, and other centres in charge of the very complicated, but very effective muscular act of coughing and sneezing. The act of blinking is also a performance attended to without our knowledge or attention. Sometimes certain nerve centres will be made active by causes unknown to us, or by fear, anger, joy, etc., and cause a person to act in a different manner. The nerve centres in charge of the heart sometimes by reason of fear or anger, will cause the heart to beat very rapidly. Nerve centres in some parts of the spine or brain will get control and direct the activities of the brain. The nerve centres in charge of balancing, walking, etc., are located in the spinal cord. These acts are very complicated and to be able to direct the execution by the muscles of the very complicated and difficult acts with speed and dexterity, certainly requires great intelligence.

The nerve centres will look after their business even if the animal's head is cut off. For instance, a dog with his head cut off or his brain removed will try to scratch the place on his shoulder which you irritate in imitation of a flee bite. In the same way a frog with his head cut off will scratch away from his body a piece of paper with acid on it, which is burning the skin.

These experiments show that the skin cells are alive and suffer from pain just the same after the head is cut off as before; that the nerve centres, whose business it is to order and direct the scratching, take charge and compel the leg to scratch the irritation or pain away from that particular spot. Now this is very conclusive proof that conscious intelligence exists in each and every separate cell of the body. How can that dog's leg be made to scratch that spot, unless directed to do so by some conscious intelligence? It would be impossible. The cells in the nerve centre or little brain of the spine of the dog, directing all those complicated muscular acts of bringing the hind leg up to the exact spot and scratching it, after having first been informed by the skin cells of the serious trouble, certainly show as much intelligence as any other act directed by the brain cells while his head was attached.

To be able to direct the scratching to the exact spot shows a will and a purpose, not only a will to do something but an ability to act so as to effect the purpose. In regard to this one movement of the headless dog and frog, Prof. Harris states: "Not only does there seem purpose in this action, but their precision as regard to place aimed at is very striking. There is no haphazard fumbling, as it were, on the part of the muscles, they answer to the message without undue delay and with all due accuracy. Now this adaption or co-ordination must depend on a very perfectly working intra-neural mechanism."

We now come to another interesting kind of intelligence we call habit. We begin practicing certain actions, as walking, riding, and playing a piano and at first every act must be directed by the brain cells, as no particular nerves have been set apart for any such special performance. But as we continue to perform these acts, special nerve centres develop and take charge of the work, until finally, after long practice the work will be performed without much, if any attention whatever, from the brain cells.

It seems to be the wish of all the cells of the body that the cells of the brain who have charge of the special senses shall not be occupied with other work more than is necessary; so as soon as is possible, when any special work is required, the organs or limbs soon learn to take charge of the work without requiring any direction to do so by the brain cells.

You can clearly see the reason for this; if the party in charge of the periscope of the submarine should have to be occupied continually with directing the performance of details which could be done by others, he could not attend to his own work, which should be obtaining information from the outside world, and directing the more important acts of the boat in its movements of capture and escape.

I started out to prove in this chapter that:

First—Any animal is an individual, made up of still smaller individuals we call cells.

Second—That the intelligence of any individual will depend on the parties in charge and directing its actions.

Third—That the parties in charge and directing the actions of an animal are those cells set apart for that purpose, which we call nervous system or nerve cells.

Fourth—That because the cells of the brain are in touch with all the other cells of the body, does not make them any more intelligent or different from the other nerve cells or nerve centres which work under them or work separately and not under them.

Fifth—That every nerve cell or nerve centre is a center of conscious and intelligent direction and directs and guides the actions of certain muscles or cells they have in charge.

Sixth—Therefore, every cell is an intelligent individual, capable of either receiving an order or executing an intelligent act, or of receiving and ordering another cell to do and perform intelligent acts.

Now it must be clear that every cell of the body is a nerve cell, in so far as it has ability to receive information together with those special qualities added which make it a bone cell, liver cell, lung cell, muscle cell, kidney cell, etc., as the case may be. That is, every cell must be an intelligent individual, in order to be able to receive orders and execute them. The cells of the liver, lungs, bones, etc., have not only special work to do, such as would pertain to the special business or work they have in charge, but they also have to receive instructions and orders and perform the work they are told to do by the nerve cells.

From a consideration of these facts, it is clear that every cell of the body must be an intelligent being. The nerve centres as directed by the higher centres in the brain and spine, are, of course, the caretakers and governors of the body. They govern the different organs, attend to the circulation of the blood, regulate the heart, take care of waste matter, direct the digestive organs, liver and every other organ.

The brain is the center of intelligence for the entire body as a whole. The following is a good description of a body or cell community, by Prof. Benjamin Moore, Professor of Chemistry in the University of London:

"The body of one of the higher animals or man consists of an enormous assemblage or community of many millions of millions of such living units far outnumbering the total population of human individuals on the earth, and this vast community of living cells which together constitute a living man or woman, are, in a state of health, so co-ordinated and regulated as to excel, in goodness of government and co-adaptation to one another's wants, any social system which has ever regulated a body corporate in human history. There is just as much division of labor and mutual assistance and government as in a state or vast empire, and moreover there are scarcely any of the defects of a bad government in the affairs of men in a social community which may not find their parallel in the organic happenings in a single human body when invaded by disease. Similar types of cell are aggregated into formations called tissues, designed for serving some common office of the body, and at times two or three tissues are blended together to form what is termed an organ for carrying out some special task. The stomach may be taken as an example of such an organ. Internally it possesses a layer in which are millions of cells formed into little tube-like glands, which secrete a digestive fluid and pour it out by millions of minute ducts or pores upon the food contained in the cavity of the stomach.

The cell must have lived for ages before it discovered the advantages of working together in a social and co-operative community, before it even started to build any structures like plants and animals. So man lived for ages before he discovered the benefits of working together in a civilized community. However, we know that savage man works and moves with a purpose and has the same intellectual machinery as civilized man and other animals. They all work for a purpose and do what they think or believe to be the right thing to do. If the cells have organized these social communities, just as man has his, why are the cells not as intelligent as man? Civilized man only a few years ago thought that burning witches was the right thing to do. Just a few years ago, you might say, the most highly civilized people in the world tied their wives, mothers and sisters to stakes and burned them alive. They did it with a purpose, thinking it was the right and proper thing to do. People are doing things today just as ridiculous and absurd. We are intelligent beings, at least we are supposed to be. Do insects ever do anything as foolish? When we investigate the wonderful organization of the cell communities which make up the body of an animal or plant, it is impossible to see anything but actions performed with a purpose and directed by intelligence. Take for instance, in the case of an infant, why is the baby born so helpless? Why is he not born like the pig, with ability to care for himself and run like a rabbit within twenty-four hours? When you come to consider the matter, however, he is able to do everything necessary for his existence. He can make a noise, breath and suck. He knows how to suck as if he had practiced it for years.

The cells of the body which have charge of the work of building and starting a new human being in the world, take care of the business at every step; as soon as he gets too large to be taken care of inside of the body, elaborate preparations take place for disconnecting him with the inside conditions. Food is prepared for him by a force of cells at the place where it is to be delivered to him after he gets outside.

This food is so properly proportioned and balanced that he will get everything necessary for the further development of his body. Some one must be interested in his welfare, and take care of him with the same faith and good will as before, so the cells that had charge of him while inside of the body continue to look after him, and these cells are allowed full control of the brain cells and all other cells of the mother. Her mind is taken possession of by these cells or nerve centres who have the work in hand of perpetuating the race and she is compelled to think that her infant is the only thing in the world worth while. She will sacrifice her life and everything, if necessary, to save her baby.

We find upon investigation that nerve cells have been placed near the lips and the mouth of the infant, which direct every muscular act of the infant's mouth in the act of sucking. We have found now that the infant's muscles are able to walk when he is born, but he has not the nerve centres developed to direct the muscular action required in the act of walking; that as soon as they are developed, he will walk whether he has had any practice or not. How the infant's mouth can know how to suck and his feet to walk, will be explained under the chapter on inheritance.

Some decline to consider the idea that anything as small as a cell can have intelligence. How about the ant? He lives a social life in the same manner as the most highly civilized species of man.

He keeps captive certain bugs that secrete a sweet juice for him, on which he feeds, just as we keep cows. He keeps slaves, warriors, servants, has grave yards, performs funeral services and many other social acts and customs performed by man. The ant's brain, where the center of his intelligence is located, is composed of but few cells,—so few that the whole crowd of cells in charge of all the social and various actions of the ant can be seen only with the microscope.

Again the actions of a louse, which lives under the mouth of the ant, exhibits just as much intelligence as a man would show placed under similar circumstances. The cell which builds the ant-louse, the ant, the animal and man are one and the same kind of individual. There are no features of distinction about any of them, so that you can tell one from the other, when examined by the microscope. The only difference that can exist in their make up is a difference in their experience. One kind has been living one kind of life and one another.

One understands how to build an ant because he has had experience in that work and in that kind of existence and none other; the other kind has had experience in building the individual we call man. In reference to the intelligent acts of the ants and the ant-louse, I must quote Prof. C. E. Walker, of the Royal College of Science in London:

"A mite, Antennophorus uhlmanni, is parasitic upon an ant (Lasius). The mite is of such a size that it would hamper the movements of the ant if it were attached on one side or the other. It attaches itself under the middle of the ant's head. The mite eats the food of the ant, but does not injure it individually; for anything that prevented the ant seeking its food would directly injure the parasite. When the ant feeds, the mite moves forward slightly, and takes its share of food from the ant's tongue. Should a second mite attach itself to an ant, the first moves over to one side and the second takes up a position upon the opposite side, thus keeping the ant well balanced and hampering its movements as little as possible. Should one of the mites drop off, the other moves to the middle. When three mites attach themselves to one ant they arrange themselves one on either side and one in the middle, and if one drops off the remaining two arrange themselves accordingly. Are these instinctive actions conceivable as the outcome of a mutation? Hardly, unless of a long series of minute mutations indistinguishable from fluctuating variations.

"Among the ants we find, besides species that live in colonies in an independent manner, species that make slaves of other species of ants. Among these slave owners we find gradations between those which are often without slaves and perfectly able to look after themselves and their young, and those which are not able to feed themselves or even to fight. Formica sanguinea is the only slave-making ant in Great Britain. Regular expeditions are made in which the nests of other ants are attacked, and the pupae are carried off by the conquerors. When the pupae hatch out they do most of the housework for their owners, who are very lazy. There are workers, however, who do the housework duties in the absence of the slaves. Polyergus rufescens has gone further in the slave-owning direction than F. sanguinea. It is admirably adapted to its part. While F. sanguinea is not really a great warrior, and fights only in numbers, P. rufescens does not mind any odds, and fights single-handed most splendidly. When in an attacking force, however, these ants act in concert in the most remarkable manner, the fierceness of the individual being subordinated to the various necessary tactical movements, even when these involve retreat. Their mandibles are so modified from the usual form that they are practically useless to the owner except for fighting. The ant is thus not only unable to perform the delicate actions involved in feeding its own larvae, but it is quite unable to feed itself, and will starve in the midst of plenty unless its slaves are there to feed it. P. Rufescens attacks the nests of other ants, killing many of them, and carrying off some of the larger pupae. When these hatch out, they serve their masters willingly and faithfully, even fighting in defense of the nest. In an allied American species, the slaves even carry their masters and mistresses when the colony migrate, as the latter do nothing for themselves. The extreme of dependence is exhibited by Anergates, which is unable to feed itself or its larvae, and would be quite unable to cope with other ants that live with and tend it, did the latter decline to serve."

The size of an animal has nothing to do with its intelligence. Compare for instance the intelligence of the insects with that of man. The insects do the same thinking and planning that man does. The brain of the insects is but a microscopic speck. The bees work together for a purpose, like the organization of a large city or factory. The ants have their slaves, and keep cows and milk them. Some ants are parasites and live upon others as men do.

It is now admitted by the scientific world that the insect known as the Hunter Wasp is able to perform surgical operations which would puzzle the greatest scientists of today. He is able to do so without any previous experience or instruction in the art. The Hunter Wasp catches the caterpillar or cricket and stings it in such a way as to paralize the creature completely, preventing all movement and yet not destroying life. The insects like the caterpillar, cricket, etc., have a nervous system which is so well understood by the Hunter Wasp that he stings it in the exact nerve center which will paralize, but not kill. In this way the wasp provides fresh meat for its young. The captured paralized victim is carried to the nest, and the wasp lays its egg on a tender spot on the paralized prey. No skilled surgeon can equal the wasp in suspending animation, and keeping the victim alive. It is now conceded that if the wasp's victim was kept alive without food or drink he would live but a few days, whereas that same creature paralized by the wasp's poisonous sting will live for several weeks. Where does man exhibit any more foresight and intelligence than does this wasp?

The mother wasp provides for a generation to come, which she has never seen. Some wise men do the same thing. The young wasp who has never seen its mother, nor had any experience or instructions in the stinging and paralyzing business, starts out and does the same thing over again that its mother did. It knows how to use its sting to paralyze the cricket without destroying its life; it knows where to find them; it knows where to find the material with which to build its nest; it performs no foolish, silly and useless actions. Man,—that is some men, show intelligence of a very high degree similar to the insects; but consider for a moment the countless absurd and silly actions of the great majority of men today. I shall not mention the every-day foolish actions that take place, but shall call your attention to some, in the past, the consideration of which will not be opposed by the prejudices of the reader. I read in the Mason City Daily Times the following:

"'Festival of the Ass.' If you should enter the church today and find an ass tied near the altar; and if, at the conclusion of the services, the priest or preacher should bray three times, instead of the usual benediction, and the congregation respond with a general 'hee-hawing' wouldn't you immediately jump to the conclusion that you were in the chapel of a lunatic asylum? Probably so. Yet, what you had witnessed and heard might be nothing less innocent than the revival of the ancient 'Festival of the Ass,' which for centuries was celebrated annually on the fourteenth day of January in churches throughout Europe. The assinine fiesta was one of the most popular of the year, and, although it now seems ridiculous, it was looked upon as a very solemn and sanctified occasion in the days of our remote ancestors."

Nowadays the dictionaries define an ass as both quadruped and biped, the latter being a 'dull, heavy and stupid fellow, a dolt, a fool.' In the popular imagination an ass is considered about the lowest in intelligence of the animal kingdom. As a matter of fact, there is no more reason for condemning the ass as lacking in intelligence than there is for praising the wisdom of the owl.

The ass has long played a prominent part in religious history. The celebrated ass of Balaam, as all students of scripture know, was wiser than his master. In the pagan mythology the ass held an honored place and it was believed that an ass, by its braying, had saved Vesta from violation. In the ancient world the 'coronation of the ass' formed an important part of the feast of the chaste goddess.

The observance of the Christian "Festival of the Ass" on the fourteenth of January spread over Europe and for centuries the feast held a prominent place in the Christian Calendar. The animal was decorated with gold embroidered trappings and was looked upon as sacred. After the procession had wended its way through the streets of the city or town, the marchers chanted a chorus interspersed with loud 'hee-haws.' The ass was a matter of much concern, as it was desirable to secure a gentle and obedient animal. The ass is noted for his perversity and obstinacy, and these qualities were often in evidence during the processions. Often the 'sacred ass' manifested a decided objection to entering the church, and in many cases had to be carried in—a proceeding which detracted from the dignity of the occasion. It was considered a great triumph, and a good omen, if the ass could be induced to bray during the church services."

You notice the ass generally refused to go in and preach, but when he did go in and stood in front of the altar and brayed, they considered the noise a wonderful oration. The ass could see no use or purpose in the fool business, and sometimes had to be carried in. However even after they had carried him in up to the altar, he would sometimes refuse to bray or preach for them, when of course they would feel greatly disappointed. This is only one of the countless absurdities performed by man and the old notion that man alone reasons and animals act from instinct is clearly without foundation. Animals never act silly as the human race does today. The oyster building cells can make a better pearl than the greatest scientist today. The oyster in his ways and place in life is just as smart as the brightest man in his place in nature. The size of the being or cell colony does not determine its intelligence, capacity or wisdom.

The Dinosaur, one of the largest animals that ever lived, is now extinct and no more. In its day it was probably the most powerful animal that ever lived. The cells building the Dinosaur went too far in the matter of size, and failed altogether while the smaller cell colonies like plants, insects, etc., are still with us and flourish in great profusion. How can the young wasp know how to do all this difficult work without any previous experience or instruction? The cells that build the young or new wasp came from the mother wasp. It is the cells that know how, not the wasp, just as it is the people in the battleship that know how to shoot and not the battleship. Just so, it is the cells in your head that know how to do a thing and not yourself; and every specific part of your knowledge is in charge of some certain bunch of cells in your brain, who have that specific department of knowledge in charge. It is likely that the wasp building cells experimented for hundreds and thousands of years before they completely mastered the profession of stinging the insects just in the right place to effect the desired results. The cells in the new wasp must be in possession of the same knowledge as the mother as they are of one and the same family.

Compare the actions of the Hunter Wasp whose brain is composed of possibly only a half dozen individual cells, whose brain is but an invisible microscopic speck, with the actions of man, whose brain is composed of more than 800 million individual cells. Whose actions show more intelligence, those of the wasp as a skilled surgeon, or those of the men trying to make the ass make a fool of himself. by putting him up in front of the altar to bray?

The brain of man is at least a million times larger than the brain of the wasp, and still its capacity to reason is not nearly as good as that of the wasp. The actions of all animals, insects and plants when closely investigated show no foolish and purposeless actions, like man's. This is clear proof that the cell as a separate individual is just as intelligent as a billion individuals. It is clear that if the wasp building cells know how to make a special instrument or stinger with which to be able to pierce and inject poison into the hard protected body of other insects, they should also know how to use it when they are ready to do so. The cells build the wasp with his stinger, eyes and limbs, with a purpose in view, just as we build a battleship with a purpose in view, and when we have it ready we know how to use it. The actions of every wing, leg, etc., of the wasp, are directed by cells in his body, who have those specific actions in charge. So it is with our actions, and so it is with the actions of the battleship. The battleship can talk, just as we can, but the actions of blowing the whistle and interpreting the meaning, must be done by the individuals in charge.

You can say that the battleship New York signalled the battleship Minnesota to do this and to do that, and in a certain sense it is true that the battleships as individuals talk to each other and direct each other's actions. In a similar manner, animals and man talk to each other and direct each other's actions, but with man and animals as well as with the battleships, it is the individuals in charge that signal and talk and direct the actions. In man, animals, and plants, it is the cell that does the signalling, talking and directing, while in the case of the battleships, railroads, cities, states, and nations, the signalling, talking and directing is done by man. All the mysteries of life become clear when we understand what it is.

First we have the primordial cell, which at this time we can hardly see with the most powerful microscope. These have organized themselves together into a larger individual or colony, whose actions we can clearly see and which we call a cell. These cells have again organized themselves into larger individuals or colonies which we call animals and plants, depending on their method and place in life. Then we have a certain species of these cell colonies or individuals we call men, organizing themselves into individuals like battleships, cities, states and nations. You will notice that one organization is based upon the other, the larger on the smaller. At the base of all is the barely visible primordial cell, who is organized into the cells that build us and all living things we see. It is the intelligence of man that directs the actions of battleships, railroad trains, cities and nations, and it is the intelligence of the cell that directs the actions of man, animals, and plants.

I could go on indefinitely illustrating these matters, that the size of brain or number of cells in the brain does not increase the intelligence. The increase in numbers gives the power to specialize in more different kinds of thinking. It is clear that every line of thinking must have its special committee of thinkers to take care of it. But as the thinkers are cells and one can think as well as the other, it is clear that the business in hand will be taken care of as well in one case as in the other, whether it is one cell or a million. The intelligent acts of the cell living its single and separate life, clearly proves that fact. I believe that I have demonstrated what must be understood by intelligence. It is a quality based on memory and possessed by the individual, who builds all organic life, the being we call cell. So far we can clearly see and understand life, at this time. Some day we shall understand it still better.

CHAPTER VI.

INTELLIGENCE OF THE CELL.

In a previous chapter we discussed the question as to what was intelligence in an animal. We found it consisted of the work of two departments of the individual, the sense organs and the brain. The sense organs must gather the information from the outside world and transmit it to the cells in the brain and the brain cells must act on such information. These are the requirements for the performance of an intelligent act by an animal, based on every other intelligent act and power, which we call memory. Memory is the power to take and keep a record of past events and use it as a reference and guide to future acts.

This power of storing away memoranda of different transactions that have taken place in the past, we find is possessed by all cells or living beings.

Three things are necessary to make up the mental machinery of an individual, viz., to receive, to think and to direct. Those three things go together to make up what we call the mind.

In the past the subject of mind had been studied as the human mind, animal mind and child mind, but of late it has been recognized that all living beings have a mind. Now this question of mind can be studied in two ways; first, by examining your own mind and the actions arising from it; and secondly, by observing the actions of others.

From late investigations, it has become clear that the mind of man is the result of the minds of the individual cells working together in his head, which we call in the aggregate, his brain.

The real thinkers are the brain cells. They are there for that special purpose. The minds of men are not all alike because they have not all received the same information from the outside world. The cells of the brain can only act on such information as they get from the outside world. If they are told for a number of years that the Catholic religion is the best it will take considerable proof to change them to Mohammedans, and just as difficult will it be to change a Mohammedan into a Catholic.

Some one has said that the greatest thing in the world is mind. It is true the mind in man has done all those things that man has produced, such as building railroads, and other industries of commerce, literature, art, science, social institutions and government. We can say further that the mind of the cell has produced all those wonderful structures we call plants and animals.

The mind has been generally considered from two points of view, as thought and conduct, that is to say, first as receiving information and learning, and then as acting, willing or doing. I think it would be more correct to divide it into three departments; first, the receiving of information from the outside world, such as is performed by the senses; second, the consideration of such information, such as the work of the purely thinking cells of the brain; third, the willing and doing department or motor brain cells, or motor system, which commands and directs the actions.

Now in order that the thinkers may be able to act on any information received, they must of course be able to use their power of memory—that is to refer to past records and experiences, so as to be able to reason out the best thing to be done at all times, as information comes in.

Let us consider this matter of conduct, in regard to determining a person's mind. We know that every being acts according to his mind. So from the actions of any individual, we can determine his mental condition, intellectual or mental capacity. We know this to be a fact, so when we see some good piece of machinery or some fine literary production or some beautiful structure or an efficient organization, we are forced to admit that mind and intellect were necessary for its production.

Before discussing any specific action of the cell, I shall quote Dr. Swoboda's description of the cell as an organizer and builder of the human body, to-wit:

"The human organism or body is developed and organized out of one cell. This original cell through sub-division and growth creates billions of cells of different varieties and character, and thus the body is but a nation or vast army of individuals; these individuals, the cells, live together for a selfish purpose, while at the same time having the community interest, just like man who is a member of a nation or community.

"Originally cells evidently existed in an environment which made it possible for them to live and perform all their functions of life individually without the aid of cooperation and organization, but conditions generally and persistently changed so that the cells were compelled to unite into organizations which were more aggressive and capable of self-preservation accordingly as the various experiences indicated and demanded, and in harmony with their capacity to organize under the influence of conscious energy, the designer, organizer, evolutionizer and life of the body.

"In a community the complete function of life is subdivided. Some individuals perform one kind of labor while other special duties are assumed by individuals who have especially adapted themselves to the requirements of the specific labor function. For instance, there are those who are engaged in banking, while others are engaged in the transportation of matter from one community to another; there are men who make storage their business and life function, while some raise cattle and thus provide the meat for the rest of the individuals. Then there are the doctors, the lawyers, and the many specialists who perform peculiar labors which are essential to the happiness and success of the community, and from this collection of individuals are chosen men who pass the laws and govern the citizens.

"Identically this same organization and specialization of the units prevails in the body. All cells have developed, to an unusually high degree, some special characteristic under the influence of conscious energy, so as to perform a special labor in the body for the benefit of the community cells, in return for which they are supplied with the necessaries of life just as the farmer has his clothes made by another, his coal dug by still another and his implements manufactured by those who have specialized in these matters and who, in turn, receive the benefit of the farmer's labor.

"As I said before, in the body as well as in the civilized state, community or city, there is co-operation, organization and specialization of the individuals. There are the liver cells, muscle cells, kidney cells, bone cells, digestive-system cells, nerve cells, and many other cells of which all of the organs and tissues are composed. Thus it is clearly seen that the entire community of cells, or the body, through all of its organs, performs the labor necessary to its survival or the self-preservation of the cells in harmony with, and in spite of, the many adverse conditions of nature at the present age.

"The brain and nerve cells are the individuals who attend to the desires and the needs of the different cells, as well as of the body, by informing the different localities of their duties. In other words, the nerve system is the governor of the body; it is made up of thousands of small nerve centers and a few large collections of nerve centers.

"The thinking center is located in the surface of that side of the brain, which controls the more active or stronger side of the body."

This is a good description of the activities of the cells in the body. If the cell is the designer, organizer and evolutionizer of the human body, as he states he is, every one must admit that he is the one that does all the business, and I for one am compelled to say that he is an intelligent being. If they can organize, design and live together in a social community, for their own selfish purpose, in a moving structure like a human being, they must be intelligent.

Did you ever stop to think of the enormous amount of skill and knowledge necessary to maintain and repair the body? There is no work in the development of organic life that requires such accurate knowledge and faithful execution at all times, as does the work of keeping the body in repair. This will also include disposing of waste material and worn out parts. This work is taken care of without the knowledge of the upper brain cells. Disease bacteria or germs of all kinds are everywhere, watching for the slightest opportunity to enter the body. These lodge in the throat, nose and mouth and are known as a cold or catarrh or pneumonia germs. They must be destroyed before they multiply and get into the blood. Who looks after this work?

The cells of the body, which we call the white cells, are cells that have not taken upon themselves any particular work, like the cells of the muscles and nerves, but live as separate beings in the body in the same manner as the amoeba now lives in water. These cells have the work of destroying invading armies of other cells, such as disease bacteria of all kinds, and also of repairing broken parts. If you cut your finger, they will rush to the spot in countless numbers and commence at once to close up the cut. To do this they will sacrifice themselves, if necessary, in destroying and fighting germs, trying to enter the body through the cut. In the struggle for existence it is necessary at times under certain circumstances for one individual to sacrifice his life for others. It is done by an intelligent being exercising his intelligence and judgment in the matter on the theory that it is the best that can be done under those particular circumstances. Here we might also consider the fact that the body has to do the best that can be done in each particular case,—for instance, if for some reason a broken bone in an animal can not be healed, it will proceed to make a joint at the place. I quote from Spencer:

"But the most remarkable modification of this order occurs in united fractures. 'False Joints' are often formed, joints which rudely simulate the hinge structure or ball and socket structure according as the muscles tend to produce a motion of flexion and extension or one of rotation. In one case according to Rotikonsky, the two ends of the broken bone became smooth and covered with periosteum and fibrous tissue, and were attached by ligaments that allowed motion, and in the other case the ends similarly clothed with the appropriate membrane became the one convex and the other concave, and were enclosed in a capsule and were even occasionally supplied with synovial fluid." Think of the broken bone, the torn blood vessel, waste matter that must be removed and new parts reconstructed. In their efforts to repair a broken limb the cells decide what is the best possible thing that can be done under the conditions of that particular case, and a hinge joint or a ball and socket joint is formed according as the conditions may require. Under these particular circumstances there is nothing done by chance. Every step requires intelligence and good judgment. The right thing must be done at the right time.

This is only one instance. Think of all the conditions and emergencies that arise during the life of an individual. Think of the "rough and tumble," ever varying game of life, and how it is necessary that the builders of these structures be not only good builders, but that they each and all be equal to any emergency that may arise. Here is where the old idea of chance is wiped out. Think of the enormous amount of work that must be done just so, depending upon the particular circumstances of the case. In a case of repairing a crushed or broken limb, sometimes new arteries must be provided, if not, the limb further down could receive no nourishment.

When the white cells rush to the place, like a wrecking crew to a railroad wreck, and proceed to clear away the wreckage and build it back into a useable condition, every act must be done with a purpose, to effect certain ends. Every move must be intelligent, just as in the taking care of a railroad wreck. The correct size of the artery and other blood vessels must be determined upon, proper materials provided, and so on in every detail of the work.

How are these beings able to communicate to each other what each shall do in these cases? We do not know their language any more than we know the language of bees and ants, who also live a social life, like civilized man. Mr. Binet has the following to say about the actions of certain cells:

"Infusoria placed in a preparation where they are at their ease, swim quietly about; if any sharp excitation disturbs them, they accelerate their pace; those armed with a rigid bristle at the posterior extremity rush precipitately onward whenever another infusory chances to touch that tactile appendage.

"It is not known whether there are many animalcula that perceive the presence of nutriment from a distance and without coming in direct contact with it; it appears, however, that this is the case with the Didinium which shatters its prey from a distance and without touching it."

Can you conceive those actions here described by Mr. Binet, as being anything but the ordinary intelligent actions of animals that you know? He describes the "Didinium" as being able to kill its victim at a distance. Can you conceive of any being who is able to make a gun, go hunting and kill its victim as not being intelligent?

Those are acts performed by the cell that lives singly and separately in the water, taking care of himself in the best way he can.

A text book on Physiology describes the cell which does our thinking as follows:

"Like the functions of all other organs, those of the brain are effected by the cells, which make up the organ. These brain-cells, which are also known as soul-cells, ganglionic cells, or neurona, are real nucleated cells of a very elaborate structure. It is true that the senses are the original source of all knowledge; but, in order to have real knowledge and thought the specific task of reason, the impressions received from the external world by the sense-organs and their nerves and centers must be combined in the association-centers and elaborated in the conscious thought-centers. The conducting path which unites a sense-organ with the cerebral cortex, or the latter with a muscle, appears as a chain of living individuals of which every member, although always dependent upon its neighbors, still leads a separate life, the specific character of which is generally different in the different parts of the nervous system."

This description goes on further and shows how the brain is simply groups of cells, each group having charge of their particular department of the work; and that disease or destruction of a particular place or part of the brain, destroys a particular faculty. It states:

"Thus disease of the speech-centre, in the third frontal convolution, destroys the power of speech; the destruction of the visual region (in the occipital convolutions) does away with the power of sight; the lesion of the temporal convolutions destroys hearing. When cells in charge of the work are destroyed the business stops."

Mr. Haeckel makes the following statement in reference to the basis of morality. He goes on to show how the idea of morality was first conceived by the cells, when they began to work together for the common good of all, that is, when they began to associate themselves into communities, and with their united efforts produced those structures we call plants and animals. He says:

"We find, even among these unicellulars (first protophyta, then protozoa), the important principle which lies at the base of morality, association, or the formation of communities. The adaptation of the united cell-individuals to each other and to the common environment is the physiological foundation of the first traces of morality among the protists. All the unicellulars that abandon their isolated eremitic lives and unite to form communities, are compelled to restrict their natural egoism, and make concessions to altruism in the common interest. Even in the globular coenobia of volvox and magosphaera the special form and movement and mode of reproduction are determined by the compromise between the egoistic instincts of the individual cells and the altruistic need of the community."

This great German scientist has got his mind so filled up with the old ideas of what is understood by intelligent beings, that he can not get his mind to work into seeing the possibility of a microscopic being like a cell possessing intelligence. He thinks that his size is against him. He must admit that the cell has had enough intelligence to build up the intelligent individual known as Mr. Haeckel, and still Mr. Haeckel will claim the intelligence himself and deny it to his maker. The following description of a single cell and of those just beginning to associate themselves together in communities like plants and animals, is from a textbook on zoology:

"In some cells, special parts are covered with countless hairs or cilia, which strike the water in a uniform direction, like a row of oars, and force the animal forward. Thus we see that the single cell is capable of very different adaptations and so we can not be surprised if the cells that compose the higher animals assume such enormously different forms. In the protozoa, the one cell discharges all the vital functions of locomotion, nutrition, respiration and reproduction.

"There is a certain animal in our fresh waters called the Pandorina. It consists of sixteen cells, all homogeneous, and each discharging all the functions. Each can produce the animal by detaching itself from the cluster and subdividing until it makes sixteen cells. Here is no division of labor. As the first step in the division of labor, we might suggest the volvox, a green globule about the size of a pin head, consisting of a number of cells. Most of the cells in this animal have taken charge of nutrition and locomotion and a few others of reproduction."

You see it is just as natural for these individuals to see the benefit of specializing, as it is for us to let each one or each crowd look after his particular line of work. We never stop to consider all the different matters to be looked after in maintaining a plant or animal. There is the simple matter of keeping the animal at even temperature, regardless of the temperature of the surrounding atmosphere; there must be some method to retain or drive out the heat from the plant or animal. The following article by Mr. Bastin in the Scientific American is interesting:

"In the first place the process of respiration is associated with considerable liberation of heat, as in germinating seeds, which show a difference of 2 degrees Fahr. in peas. The respiration of the seeds involves the taking in of oxygen and the giving out of carbon dioxide, the same process as that in animals. It is the habit of the Alpine plant, Soldanmillas, to start active growth early in the Spring before the snow goes away, yet owing to the large amount of heat which is liberated by the shooting flower, these are able to bear their way up through the snow. The flower stalks are very slender which is simply owing to the heat they liberate that they can make their way through the snow crust which is generally exceedingly hard. Where the snow is very deep, they do not reach the surface at all but expand in a cavity which they have thawed out in shape of a dome where they blossom and develope their pollen.

"Experiments have shown that inside the spathe of the

Brazilian Delkciosa, the temperature was 100 degrees Fahr., while the air outside was 78 degrees Fahr. Even more astonishing was the case of the Arum Italicum. Here, when the outside air was registered at 60 degrees Fahr. the inside of the spathe showed a temperature of 110 degrees Fahr. The temperature of a plant, like that of an animal, is a fairly stable feature. No matter what the external conditions may be, the ordinary temperature of the human being remains at about 98 degrees Fahr., whether living in the tropical or arctic regions. It is the same with plants; the long spells of hot dry weather in Australia are notorious, but however scorching the sun may be the sap of the plants remains cool. The long roots of the Eucalyptus tree are tapped for their cool sap by the natives. The root 20 to 30 feet running in 6 inches of baking soil, is cool and delicious. Melons and gourds of all kinds are astonishing in their powers of keeping cool, even under the most scorching conditions. In the testing of the temperature in a fairly strong sun, it ran up to 110 degrees Fahr., yet the temperature in the interior of the fruit was little more than 60 degrees. The maintaining of the temperature in the plant is a process which is directly associated with the life of the specimen. The coolness of the gourd remains only so long as it is attached to the plant. When the fruit is cut it rapidly becomes the same temperature as the atmosphere, like any other object. The moisture in the plant can not be kept cool by evaporation, as it can spare no water but must conserve every drop. Everything we know about a plant which lives in a dry hot climate shows that all kinds of devices are employed to check and prevent loss of moisture. The loss of moisture by evaporation through the thick skin of the gourd is very slight, while the leathery skin of the others, like the cacti, hardly transpires at all."

To keep the human body in an even temperature and to keep the plant warm enough to melt the snow around it, is a continuous job, requiring great skill. The cells know how to create heat and in the same way the southern plant knows how to create cold. They obtain the results but in different ways.

In the previous article from the school physiology, we noticed that the cells in the brain although they were connected together in a mass, still retained their individual separate existence, so each one must have a mind of his own. Still, all working together, they form the mind of the individual in the same way as all the combined will of a nation or body of people becomes its will; as for instance we speak of the will of the Legislature, or the sentiment or mind of the council, public, or the German people. There is the separate mind of the individual cell and the combined mind of the body or brain. The human mind or brain is generally considered superior to all others, by reason of its size. Mr. Haeckel has the following to say about the brain:

"With the most improved means of modern histology, the most perfect microscopes and coloring methods, we are only just beginning to penetrate into the marvelous structure of the phronetal cells and their complicated grouping. Yet we have advanced far enough to regard it as the most perfect piece of cell-machinery and the highest product of organic evolution. Millions of highly differentiated phronetal cells form the several stations of this telegraph system, and thousands of millions of the finest nerve-fibrils represent the wires which connect the stations with one another and with the sense-centres on the one hand, and with the motor centres on the other.

Comparative anatomy, moreover, acquaints us with the long and gradual development which the phronema has undergone within the higher class of the vertebrates, from the amphibia and reptiles up to the birds and mammals, and, within the last class, from the monotremes and marsupials up to the apes and men. The human brain seems to us today to be the greatest marvel that plasm, or the 'living substance,' has produced in the course of millions of years."

You notice here how he states that the human brain is the greatest marvel that "plasm" or "living substance" has produced in the course of millions of years. Why not call them cells or beings, instead of such nonsense as "plasm" and "living substance." Those words are not only meaningless but misleading.

There is no difference whatever in the brain of the jackass and the brain of the President of the United States, except that one contains more cells than the other. The cells are themselves the same kind and size, but the human brain contains more of the individual cells, which gives the human mind or brain more chance or power to specialize and to think in more ways and on a larger number of subjects.

There is no question but that this extra size of the brain has weakened the power of the human mind to reason and think in a general way. I cut the following article from the Kansas City Daily, not long ago, which should give one some food for thought:

"Ten years in the federal prison at Atlanta was the penalty inflicted upon four young Navajo Indians of Arizona, who murdered a medicine man, Gi Shin Beta, because they believed the arts he practiced upon the sick 'were shot with witchcraft born of the devil,' resulting in the death of two patients who never recovered from the uncanny spell cast over them. The murder was committed by order of the chief without knowledge or fear of the white man's law and in the name of the Great Spirit, says the Fort Worth Star-Telegram.

"Natani Bezab, Hoskay Oho Neeth, Natch We No and Bitsie No Dozio were found guilty after a trial fraught with difficulty and delay. The defendants never had learned even a smattering of the English language, making it necessary to translate every question and answer. The jury consisted of white men. Bitsie No Dozio did not actively participate in the crime and was told by the court that a parole or pardon would be recommended for him within a short time.

"A young woman of the Navajo tribe was afflicted with tuberculosis. Her parents sent a messenger for Gi Shin Beta, who had a reputation for healing the sick and casting out devils. He had also a black spot on his long record. Four years ago he attended a young woman ill of a similar disease and was said to have cast a spell over her from which she never recovered.

"The girl was found in a dying condition. The four men charged with murder waited without while the treatment was being administered. The medicine man went out and consulted the stars when he noted her condition. Returning he said that he had seen a large wheel in the heavens and two men shoveling ashes into the center of it. This meant that there was no hope for her recovery.

"After she died a black spot was found over her heart, and her father instantly divined that the medicine man had shot an arrow of witchcraft through her body, thereby causing her death. He consulted his friends and the belief became general in the Indian camps that Gi Shin Beta possessed a 'body full of devils.'

"In the former case he had cast over the young woman a spell that lasted for many days. He was importuned to take it away, but he steadfastly refused, holding that he had no power to do it.

"Another medicine man was called and after examining the girl, he confirmed the beliefs of other Indians that, could the spell be removed the girl would recover. This was recalled and Gi Shin Beta was informally sentenced to die. Conforming to an unwritten law of the tribe and in the belief that they were doing their fellowmen a justice, the four young Indian braves went to the mountain lodge of Gi Shin Beta, took him from the house and killed him with an axe. They left the body in the yard and there it remained for several hours. A friend of the four came along later and rolled it into a ditch.

"The plea of H. H. Linney, council for the defendants, was unique. The defendants, he said, knew not what they did. They were following a wild call of other generations of their kind,—not the Old Testament call of an eye for an eye and a tooth for a tooth, but the call that rings in behalf of society and the peaceable livlihood of those interested in the welfare of their kin and kind. Linney declared that the white man is inconsistent in condemning a belief in witchcraft, as it has been but a few years comparatively since white men were burned at the stake by their own kind for the practice of witchcraft.

"There are thousands of men today who believe that humans are possessed of the power to cast out devils, Kinney argued, and the white man should not be too prone to condemn the Indian for holding such a belief. Religious belief is strong in all mankind, he said, and these four slayers could scarcely, in justice to their belief, be measured by the tape of the white man's law.

"There is consternation in the camps of the Navajos.

No argument that their simple minds can conceive justifies the white man in depriving members of their band of their liberty for doing their duty as patriotic representatives of a dwindling tribe."

What right have we to interfere with the religious beliefs of these people, when we believe in the same creed ourselves? Christ believed in it and practiced it, and we believe in Christ. Only a few years ago, we, the intelligent citizens of America, burned our mothers and sisters, in the belief that they were possessed of the devil.

In accordance with the Bible quotation, "Thou shall not suffer a witch to live," thousands of women were burned alive. Such highbrows as preachers, judges and statesmen were the people who directed the work, thinking that it was the best thing and the right thing to be done. Three-fourths of the people of the United States are just as silly today as they were then. What has man to boast of, in the way of reasoning powers?

We find no such preposterous and foolish actions in any animals as we do in man. There is of course practically no difference in the mind of the animals and man. This fact has now been well established on the question as to the origin of the human faculty. Mr. Romanes states, in substance as follows:

"All the emotions such as fear, surprise, affection, pugnacity, curiosity, jealousy, anger, play, sympathy, emulation, pride, resentment, emotion of the beautiful, grief, hate, cruelty. benevolence. and the emotion of the ludicrous, are found in the animal mind."

This list exhausts all the human emotions, with the exception of those which refer to religion, moral sense and perception of the sublime. Romanes states that: "The emotional life of animals is so strikingly similar to the emotional life of man, and especially young children, that I think the similarity ought fairly to be taken as direct evidence of a genetic continuity between them.

"And so it is with regard to instincts,—such as have reference to nutrition, self preservation, reproduction and rearing of progeny."

No one will venture to dispute the fact that all these instincts are identical with those which we observe in the lower animals. Language has arisen from sign making, etc. As man has come up thru the same stages of development as the other animals, and as only a few centuries ago he lived the life of an animal as animals live today, it is only natural that his thinking organization would be on the same plan, and that he would have the same nerve centres, which would produce the same actions and emotions.

Haeckel, describing the brain or thinking machinery of man states: "The phronema is the organ of thought in the same sense in which we consider the eye the organ of vision, or the heart the central organ of circulation. With the destruction of the organ its function disappears.

Between these central organs of sense lie the intellectual or thought organs, the instruments of presentation and thought, judgment and consciousness, intellect and reason; they are called thought-centres, or association centres, because the various impressions received from the sense centres are associated, combined and united in harmonious thought by them."

Then in another place he makes the following statement, showing how the human mind is deceived and tangled up with all kinds of foolish ideas:

"Modern spook-seeking has no more value than mediaeval magic, babalism, astrology, necromancy, dream-interpretation, and invocation of the devil. We must put at the same stage of superstition the spiritism and occultism we find mentioned so much in modern literature. There are always thousands of credulous folk in educated countries who are taken in by the performances of the spiritualists and their media, and are ready to believe the unbelievable. Spirit-rapping, table-turning, spirit-writing, the materialization and photographing of deceased souls, find credit, not only among the uneducated masses, but even among the most cultured, and sometimes among imaginative scientists."

It is necessary to call the attention of the reader to these facts and conditions to show what the human mental machinery really is.

The conjugation of the single cells living a separate life is described by Binet, as follows:

"Nevertheless, the two Stentors continue to be united for a certain length of time by a bridge of matter, located even with the point where the contraction took place; this bridge of matter gradually grows thinner and thinner and becomes as fine as a thread. Now Gruber has observed that the two Stentors united by this bridge of protoplasm exhibit perfect harmony in their movements; they always sway in the same direction at the same time; and this harmony is necessary, because the least contrariety of motion would suffice to break the feeble bond that unites them. Moreover, their vibratile cilia beat in unison. To explain this concordance in the movements of the two animals, Gruber assumes that the entire mass of their protoplasm performs the function of a diffused nervous system, which has the effect of regulating their movements and of making them harmonize."

Just note the actions described here, and the opinion that these actions indicate a nervous system. Now it does not make any difference what they have or what you call it, we must admit that they show intelligence of a high order. We have no microscope at this time powerful enough to see the individual or primordial beings, that make up the individual we call cell, but we can clearly see the actions and methods of the cell, and I wish the human mind to understand this proposition right and acknowledge the intelligence of his maker, the cell.

Mr. Haeckel gives him a name, as if he had no more intelligence than a stone, calls him a "plasm" or "living matter." Later in his book, he gives a description of him in a different light and compares his acts of special progress, with those of man. He makes the following statement:

"We need only to glance back half a century, and compare life today with what it was then, in order to realize the progress made. If we regard the modern state as an elaborate organism (a 'social individual of the first order'), and compare its citizens to the cells of a higher tissue-animal, the difference between the state of today and the crudest family groups of savages is not less than that between a higher metazoon (such as a vertebrate) and a coenobium of protozoa.

"The progressive division of labor, on the one hand, and the centralization of society, on the other, prepare the social body for higher functions than in isolation, and proportionately increase the worth of its life. To see this more clearly, let us compare the personal and social value of life in the five chief fields of vital activity,—nutrition, reproduction, movement, sensation, and mental life."

Now if the cell has gone through the same process as man in gradually learning, discovering and taking advantage of the benefits of social life, why has he not done so by reason of his intelligence, just as man has.

One writer describes the actions of the mental machinery of man, in the following language:

"The physiologist, as a physicist, observes how a beam of light, a wave of sound, or a vibration of heat affects the organs of sensation; how they enter the nerves, are transformed into an irritation of the nerve-fibers and conducted to the brain cells. Here he loses all trace of them.

"On the other hand, he observes a spoken word coming from the mouth of a speaking person; he sees the person move his limbs, and finds these movements are caused by muscular contractions produced through motor nerves irritated by the nerve-cells of the central organs. Here again he is at his wit's end. The bridge which should lead him from the irritated sensory nerve to the irritated motor nerve, is indicated in the labyrinthian connections of the nerve-cells, but he lacks a clue to the infinitely involved processes which are interposed in this place."

Now the actions of any submarine would seem just as mysterious if we did not realize the fact that the individual itself, which we call submarine, was in charge of intelligent beings, which are directing its actions and course according to the information that it receives from the outside world.

Just think of the busy life inside of the animal or human individual. The heart and circulating system causes a continuous supply of nourishment and oxygen to every individual cell. It is the cells in the animal that must have oxygen and food, not the animal. In the same way with the submarine, it is the people that occupy it and run it that must have the food and air. Think of the thousands of orders, going to the storage tank we call the stomach, for material to repair this and that. Think of the millions of chemists, one making gastric juice, one bile, one saliva, one tears, one this and one that. Think of the millions continually carrying food and oxygen, and the millions carrying away waste matter. Think of the millions on guard and duty night and day as superintendents and questioning all suspicious characters and killing enemies. Nothing like this order and method can be equaled in the social communities of man.

We must come back again to the consideration of the ciliated cells that live single separate lives in the water. The following is a description of the actions of those cells by Mr. Binet:

"In case of the hunter ciliates proper, the mouth is constantly closed. The act of absorbing each object captured is accomplished by the process of deglutition comparable in every phase to the like process in higher animals. Furthermore, these species feed only upon living prey, which they capture and entrammel by means of their trichocysts. By this very act they exercise a choice in the selection of food.

"These hunter-Infusoria are constantly running about in quest of prey; they move rapidly hither and thither, changing their direction every moment, with the part of the body bearing the battery of trichocysts held in advance. When chance has brought them in contact with a victim, they let fly their darts and crush it; at this point of the action they go through certain manoeuvres that are prompted by a guiding will. It very seldom happens that the shattered victim remains motionless after direct collision with the mouth of its assailant. The hunter accordingly, slowly makes his way about the scene of action, turning both right and left in search of his lifeless prey. This search lasts a minute at the most, after which, if not successful in finding his victim, he starts off once more to the chase and resumes his irregular and roving course."

How do these actions compare with the actions of man

when he is hunting? Still our great German philosopher, like all other scientists, calls this being, this intelligent hunter who makes his own weapons, with which he is able to kill his victim at a distance, only a "plasm" or "living matter."

We never have time in our busy lives to stop and consider all the wonderful things which take place in the living world around us. The acts of insects are simply astonishing. I read the following in my Sunday paper the other day, which mentions a few of their performances:

"David Fairchild, the plant explorer, has discovered that the champion athlete and aeronaut of the insect world is the king grasshopper. It can jump one hundred times its own length, and has been known to sail for one thousand miles before the wind.

"The carrying power of the song of the cricket is extraordinary. There are species whose strident notes can be heard for a mile, although their bodies are scarcely more than an inch in length. The males alone are musical.

"Of all creatures in our houses, the cockroach is the most detested. Housewives may be surprised to learn that a cockroach can live five years, and that it takes a year to develope to maturity. The female lays her eggs in a horny capsule like a spectacle case, which she carries about with her until she is ready to deposit it in some suitable place. Later she returns to help her baby cockroaches out of their shells.

"The song of the cicada is the noisiest in the insect world. The seventeen-year cicada has been called the Rip Van Winkle of the insect world. From its tiny eggs there issues a creature with soft white body and mole-like front legs. It hurries to the ground and disappears beneath its surface, sometimes to the depth of twenty feet. For seventeen years it digs its way around in absolute darkness, and then comes to the surface to join in a marriage revelry of a few weeks. It is a full-fledged creature of the air, though encased still in grave clothes of parchment; but it soon splits these up the back, pulls itself out, dries its powerful wings and flies away with the whirr of an aerodrome to live but a few weeks."

Certain fish have a habit of catching insects by squirting water on them so they will fall off the plants into the water.

Squirrels gather food for winter and hide it, but can always find the graneries despite the deep snow. A rat in Siberia, the Sagonies Pica, gathers grass in the fall which it will need in the winter; and one observer states that like any farmer he spreads it out to dry in the sun, then collects it in ricks which it shelters from rain and snow.

The crested Grebe makes a raft which floats for a nest, and if you get too near it, will push it away from you by paddling, as you would a boat.

I could go on indefinitely describing these different actions of animals, showing that all animals, including man and the insects, are intelligent beings. They appear to us to be possessed of different degrees of intelligence, but the fact is that the one who may appear the most stupid may be the most skillful in the particular line of work which concerns him and his well being, in that particular place in life where he exists.

Wherein do the actions of animals that watch and pursue their prey, lay snares for it, like the spider, and devour it, differ from the actions of man, when he hunts and does the same things? Wherein do the actions by which the animal hides itself, avoids the snares laid for it, invents deceptions for its defense, differ from the actions of man, now taking place in this war? Man acts for an end, so do animals. However, we must not forget that the real actors are the cells, in the same manner that the real actors in the submarine are the men inside.

When we see a man building a house, painting a picture, or digging a ditch we know that the man is working with a desire, an idea or purpose. We can not read the internal actions of his mind but we know from our experience what the facts are, from our own mind. It is not even necessary to see the man do the work. The work itself is proof positive of an idea or purpose.

It must be admitted that every work or product based upon an idea or purpose can be caused only by intelligence and not by chance. The actions of man and animals are for a purpose. When the cell in his single state makes his dart with which to kill other cells, it is clear that he is working with an idea or purpose. When he shoots his dart at another cell, he clearly does so with a purpose and with an idea that he is going to kill and have something to eat. We see in cells of both animals and man, actions identical. They are all the same kind, and for a purpose. I shall have to describe all these different things, so the reader can determine for himself.

I am here to defend the cell, to show that he is possessed of judgment and understanding, that he does nothing blindly. His existence still with us as a single cell is proof certain that he has a mind possessed of extraordinary resources, to be able to combat cold, hunger, death and all the elements and enemies of living matter, or else he would not be in existence today.

Our limited intellect is not a competent tribunal to pass judgment on the cell and his actions,—on one who has produced all the wonderful structures that have existed in the past history of our planet. His works in the past and today are beyond our comprehension. The idea of a true intellect in anything outside of ourselves, and especially in a microscopic being seems absurd to some people. However, when we study the actions of the cell swimming in the water, or the cell in our brain doing our thinking, we are studying the same individual, but in different situations in life.

The cell in the man's brain occupies the same place as the superintendent of a large railroad, while the single cell in the water occupies the same place as the savage who still roams the forest and hunts for a living. Each one understands his work in life, and each one acts with a purpose.

A most wonderful experiment to test and prove that every cell of the body has intelligence, as well as the brain cells, can be made with the decapitated frog or headless frog. I have tried it myself several times. Prof. James describes it as follows:

"As good an instance as can be given is the often quoted instance of the decapitated frog, which can not of course see or feel, and can not consciously perform any movement. Yet if a drop of acid is placed on the lower surface of the thigh of the frog in this state, it will rub off the drop with the upper surface of the foot of the same leg; if this foot be cut off, it can not thus act. After some fruitless efforts it gives up trying in that way, seems restless, as though it was seeking some other way, and at last it makes use of the foot of the other leg and succeeds in rubbing off the acid. Notably here we have not merely contractions of muscles but combined and harmonized contractions in due sequence for a special purpose. These are actions that have all appearances of being guided by intelligence, and instigated by will, in an animal, the recognized organ of whose intelligence and will has been removed."

This instance and many similar show that will and intelligence are everywhere in the body, and that the brain is not the only place where it is found. The brain cells have been removed, but as all cells of the body are thinkers and intelligent, other cells take their place and give orders.

And as I said before it requires intelligence to obey instructions promptly, as well as to give instructions. The cells that contracted the frog's muscles and lifted the foot to the right spot and moved it from place to place, so as to remove the acid, required just as much intelligence as the cells that ordered the muscles to do it. It required knowledge, skill and intelligent execution, to be effective. I do not see how a more conclusive test could be made. These actions will also take place in other animals with the head removed. Think of the hundreds and thousands of cells involved in the act of lifting the leg and making it rub off the acid. So I claim that each and all of the cells of the body are possessed of intelligence just as the original cell which started the construction of the body.

Under our discussion in a previous chapter of what life really is, we found that it is not any of the chemical or physical forces that we are acquainted with at the present time, but that the life force is in the being we call a cell; that this life force distinguishes itself from all other forces in this, that it is able to direct and make use of the other forces in nature, like electricity, gravitation, the movements of water and wind, heat, cold, light and the different kinds of chemical action. Certain kinds of cells that are produced by the clover-plant understand how to extract the nitrogen from the air and store it away for future use. One of the greatest chemical discoveries and inventions of the age was the secret of extracting nitrogen from the air by electricity. Why does this act by man prove him to be any more intelligent than the cell, which lives in the clover-plant and performs this same act?

The most important power or knowledge that certain kinds of cells possess, is the power of making starch and different kinds of materials which we call carbohydrates and fats from the raw material of soil, air and water. All the cells which build plants have this power, but in order to do this they must have sunlight. The cells who have this power have also a speck which resembles an eye. These cells nourish themselves by making starch with sunlight, so if they are deprived of the sun's rays, they can not make their food and thus starve to death. These cells that understand how to make starch from the air, earth and water, have a special organ for this purpose, which the other cells do not have. This organ is called the chlorophyl pigment or chromatophore. These cells carry a tiny chemical laboratory with them, with which they change a crude substance into a finished product like starch which they use for food. It is a singular thing that they should use the same substance for food that we do. These cells that carry a starch factory with them have also an eye. In regard to this, one observer states:

"If flagellates possessing chromatophores, that is organs generating starch, have ocular spots at the same time, it is because these rudimentary eyes enable them to find their way towards the light, which is the necessary agent of chlorophyl action. Accordingly all micro-organisms having eyes nourish themselves as plants do. In their case, the object of the eye is to direct the performance of a vegetable function.

"In the case of the euglenae, the chromatophores are formed of small discoid plates; they are situated directly under the cuticle, so that the light can act upon them.

"In certain species of flagellata, they are exhibited under the cuticle in the form of two large plates which envelop the protoplasm like a cuirass formed of two pieces."

We have to admit that we do not know yet by what process of alchemy the crude material of earth, air and water are combined or transformed into starch, but we see that it takes place. Only to those cells that depend on that method of making a living is the secret known, and the secret is passed along as the common knowledge from generation to generation.

In late years there have been some discoveries in reference to the chemical effect of sunlight on different kinds of matter. The heat or electric energy in sunlight is evidently used by the cell to start a certain molecular activity in matter which molecular activity the cell is able to direct to form the different chemical compounds wanted. This photo chemistry, as it is sometimes called, is believed by some now to be an electric energy. I read the following interesting article in regard to it some time ago, in the Scientific American:

"The subject of Photo Chemistry is one about which comparatively little is known; while the application to ordinary photography is well understood, the theory that leads to the chemical action of light is far from being perfectly comprehended. The chemical action of sunlight, such as that shown in the bleaching process and blue printing has been known for a long time. Recent investigations have taught us that numerous compounds are sensitive to light, and convinced us that we are dealing with a mutual action between ether vibrations and chemical force, but it takes place only in special cases, as light is considered to be the vibration of ether. When the ether vibrations traverse any material, they produce two different results: First, they raise the temperature; second, they occasion chemical energy. The first is called absorption of light, the second chemical change. Gases, liquids and solids all respond to ether vibrations, such as the explosive mixture of hydrogen and chlorine, chlorine water, which gives up oxygen, and phosphorous which changes to red, or cinabar which turns to black. Now, what is the cause of these light vibrations?

"The latest authorities maintain that light vibrations are produced by electric agitations, and that in the chemical action of light we deal with phenomena not far removed from the formation and decomposition of compounds under the influence of the galvanic current.

"The action of light waves (according to latest theories) are the rapidly alternating electric fields. From these conclusions we may assume that the ultimate cause of the photo-chemical action of light lies in electric phenomena."

The most common and abundant single cell that we have in our ponds and lakes which makes its own starch or food, is the cell called euglena. This is what you might call a minute invisible green plant. These Euglenas are again eaten by other cells that do not understand how to make starch for food. The Euglena is known as the green scum of sewers, lakes, ponds, streams and muddy places. As soon as it finds itself in darkness or light too weak to aid it in its starch-making activities it moves into a better light. Euglenas are cigar shaped with a tail for a propeller, which works like a cork screw. The starch factory is carried at the front end. Under the microscope these creatures can be seen gathering together at certain places where the light is not too weak nor too strong, generally about two or three inches under the water.

They are able to distinguish between the different shades of light. They move about up and down until they get into just that degree of light which seems to them about right, and then remain as nearly stationary as possible, and while in such position, cause their starch machinery to work full speed.

It is clear that this cell can see, although his eyes have not yet been discovered. It has the motor apparatus, with which it is able to go to the place where it can obtain energy from the sun with which to run its starch factory. It has a mind with a will to go to the place when and where it is necessary to go. It has the skill to run the starch factory. It has the keen sense to distinguish and discover when it is in the right position and place where the radiant energy of the sun is not too weak or too strong for it to operate its starch machine in the proper manner.

Thus we see life is not sunlight, heat, cold nor electricity nor any other of the forces of nature, with which we are acquainted. All life that we see such as plants and animals, is caused by intelligent acts of the cell. These intelligent beings are able to take advantage of all these different forces of nature and turn them into some useful purpose for themselves. They understand how to effect chemical combinations of crude matter, like earth, air and water, so as to mould them into such structures as they need for food building material or for other defensive purposes.

The sun's heat is a wonderful kind of energy with which to effect chemical changes in matter. The cells have had a chance to experiment with this energy for millions of years and they have discovered how to set up molecular disturbances and activities in matter and to direct the actions of atoms and molecules, so as to be able to effect any particular molecular combination of matter they wish.

All plant and animal cells know how to use sunlight for making the different kinds of building material which they may need. The tree must have an outside covering so with the energy of sunlight, the cells who occupy it make what we call bark from the crude material taken from earth, air and water. Heat is required to start chemical action. The sun is usually the only source of heat at the surface of the earth, and it was the most natural thing that the cell should discover methods to utilize this power. Just think what a discovery it would be, if the human race should discover this method of making starch or other fibers from earth, air and water!

Since we have already discovered the secret of extracting free nitrogen and a number of other chemicals from the air, which was known only to a few species of cells a few years ago, why should we not in time discover this most important secret of all, how to make starch and fats from crude matter? We can not hope to learn how to do these things by merely observing the actions of the cell. Our eyes are made to see things a thousand times larger. We are living in a different world. However, it is a chemical secret and it can be discovered by experimental work. We see in this microscopic wonder who makes food for himself and also for us, the same living being as ourself. He is a composition of intelligence and matter, race and individual, life and death, past and future, all gathered together in one, just as it is in us.

We see that the matter of size makes no difference. How many ages did the cell live singly and separately in the world, in this manner, making its own food, before it learned to associate itself into communities like plants and animals? When we consider the old ruins left in the earth's crust, of all the different animal and plant structures which were at one time the abode of the cell, we get an idea of the vast experience and work of the cell, and the tremendous age of this planet.

We must now consider the actions of plants. A plant is a stationary abode, so constructed that its occupants can get out into the sunlight in the summer and make up the different building material and food that it may need. Trees generally build temporary structures for the season, called leaves, where they get the use of the sunlight. When they get a "hunch" that winter is coming, they scurry back into the tree, cut the leaves loose and close up the connections. Some plants climb up the sides of other plants and structures. They build out tendrils with which they grasp hold of other plants and attach themselves permanently.

In reference to these moving, grasping, highly sensitive plants, Darwin has this to say:

"We see how high in the scale of organization a plant may arise, when we look at one of the more perfect tendril bearers. It first places its tendrils ready for action, as a polypus places its tentacula. If the tendril be displaced, it is acted upon by the force of gravity and rights itself; it is acted on by the light and bends towards or from it, or disregards it, whichever may be most advantageous. During several days the tendrils spontaneously revolve with a steady motion. The tendril strikes some object and quickly curls around and firmly grasps it. In the course of some hours it contracts into a spire, dragging up the stem and forming an excellent spring. All movements now cease, by growth the tissues soon become wonderfully strong and durable. The tendril has done its work and has done it in an admirable manner.

"Spreading out their branches in contact with any

nearly flat surface, they develop discs—these adhere by the secretion of some cement to a wall or even to a polished surface. The rapid development of these adherent discs is one of the most remarkable peculiarities possessed by any tendril. In the first species, the tendrils resemble in shape a bird's foot, they can seize firm hold of a twig or branch."

I have myself conducted some very interesting experiments with climbing plants, which have lead me to believe they can both see and feel like an animal; in fact, I have no doubt about it, as they will change their direction and move towards and find a stake or plant. When we consider the fact, however, that cells build these partly stationary and partly moving structures, we should not be surprised at their actions. You will have to admit that anything that will build these feeling and grasping structures with the idea and purpose of climbing up the side of other structures and thereby fastening themselves permanently, and with the one purpose in view of getting into the sunlight, must be possessed of a guiding and directing mind.

You can not believe that matter alone can climb up and grasp hold of a twig, or side of a wall. They wish to climb up into the sunlight and make carbohydrates, and they do it. You might say that a plant does not possess those organs wherein will, intellect and instinctive action reside. That is to say that a plant has no brain. What is a brain but a community of cells? And what is a plant but a community of cells?

This being a fact, we should expect to find the same intelligent and purposive acts in one place as in another. Mr. Haeckel makes the following comparative statement between the social organization of man, and that of the cell:

"The original physiological independence of the cells which have combined to form tissues is more completely lost in proportion to the closeness of their combination, the complexity of their division of labor, and the differentiation and centralization of the tissue-organism. Hence the various kinds of tissues in the body of the histona behave like the various classes and professions in a state. The higher the civilization and the more varied the classes of workers, the more they are dependent on each other, and the state is centralized.

"The complicated modern state, with its remarkable achievements, may be regarded as the highest stage of individual perfection which is known to us in organic nature. But we can only understand the structure of this extremely complex 'organism of the highest order,' and its social forms and functions, when we have a sociological knowledge of the various classes that compose it, and the laws of their association and division of labor; and when we have made an anthropological study of the nature of the persons who have united, under the same laws, for the formation of a community and are distributed in its various classes. The familiar arrangement of these classes, and the settling of the rank in the mass and the governing body, show us how this complex social organism is built up step by step.

"But we have to look in the same way on the cell-state, which is made up from the separate individualities in human society or in the kingdom of the tissue-animals, or the branches in the kingdom of the tissue-plants. Their complex organism, composed of various organs and tissues, can be only understood when we are acquainted with their constituent elements, the cells, and the laws according to which these elementary organisms unite to form cell-communities and tissues, and are in turn modified in the divers organs in the division of labor."

He states that the complicated modern state with its remarkable achievements may be regarded as the highest state of individual perfection which is known to us in organic nature," and he compares the social achievements of man with those of the cell, and in fact arising from the cell. The social life lived in a plant or animal compels us to recognize among the cells a spiritual communication, similar to our own. A fair consideration of these facts compels us to admit that the cell has the same intellectual capacity as man.

The remarkable harmony and unity of action, the extraordinary division of labor, the regularity with which one group of workers will take the place of another, convinces me that there is no difference in their intelligence.

Many plants, if not all of them, can both see, feel and hear. I do not mean that they can do so to the extent that we can, nor is it necessary for their existence that they should. There has been a great amount of investigation, of late, in regard to the question of the existence of a will and consciousness in all plants. Some time ago the following appeared in Current Opinion:

"Only within recent years has systematic observation been made of the results consequent upon the division into two sexes of the conspicuous forms of plant life.

"They have eyes which see (to follow the elucidation of Royal Dixon, a student, of what he deems the human side of plants), they have mouths with which they eat and stomachs to digest their food in. The stomachs of plants are in the form of leaves; but they subserve the purpose. Plants have lungs with which they breathe and they actually drink water. They are organisms, and because they are organisms they perform the functions of such things; they actually mate.

"Plants have not always had the same manner of eating, drinking and sleeping, nor have they always had the mating customs prevailing among them at present. Plant customs and habits change as do the habits of other organisms. Hence their mating habits have been modified through the ages. Before the mating of any pair of plants occurs, there is at this stage in the evolution of many among them a brief period of what must be called courtship. The happy and gallant wooer adorns himself gorgeously with brilliant flowers, each having powdered faces, calling to his love on every breeze. He must charm or there will be no response. This, of course, refers to matings among the more developed plants.

"When we speak of flowers we rarely stop to consider just what the term means. It means not only the pistil, which contains the undeveloped seeds or ovules; the stamens with their pollen grains; but the petals, or taken together, the corolla; and lastly the calyx—all these different parts combine to form the flower. The brilliantly colored petals are really used as advertisements. The red, yellows, oranges, greens, purples and whites are flags that signal to the bees and butterflies to come and feast on the honey—and thus to fill their fuzzy backs with the pollen grains which will readily cling to the sticky pistil of the next flower they visit.

"One of the most brilliant displays of color is that of the flame azalea. It flaunts its gaudy blossoms over the mountain-sides, beckoning to the pollen-bearers to come and taste of its honey. Its flame colored flowers are produced in great profusion, and massed together, their blazing splendor gives the impression of the woods on fire.

The azalea, because of its gay blossoms, is becoming very popular as a cultivated shrub.

"Some plants do not care to have their pollen distributed, but fertilize their own flowers by dropping the pollen grains upon their own pistils. But in all such cases their children are degenerates, and only plants which are very low and unsuccessful in life use this means of fertilization. While in a very large percentage of flowering plants the male and female elements both are present in the same flower, if good healthy offspring are to be produced it is necessary for pollen to be brought from another plant, or another flower of the same plant.

"It was long ago proved that close interbreeding produces degenerates in the plant kingdom. There are very few instances among high-class plants where perfect seeds have been produced without the ovules having been fertilized in the regular way; that is, by a transference of the pollen from the male to the female flower.

"Among such plants as begonias, cucumbers, gourds, squashes and the like there are many flowers that are distinctly male or female. If for any reason the proper insects do not exist in the territory where such flowers are to be raised, the flowers may be fertilized by carrying pollen dust from the male to the female by means of a feather or a dainty brush.

"Plants have various devices for securing a cross-fertilization of their flowers. Some use the wind as an agent, others depend upon the bees, the butterflies, the moths, the snails or even the birds. Bird pollenation is a common thing in Brazil, where the abuntilon is fertilized by the humming bird. Flowers use their beauty, perfume and conspicuousness to attract to themselves insects that will distribute their pollen.

"And in considering this plant courtship and marriage

there is one point which needs especial emphasis—a point which must necessarily be reiterated time and again in the consideration of the human side of plants. It is the existence of some guiding force, too impulsive to be mechanical, too versatile and efficient to be instinctive, which controls the actions and manners of plants in all the stages of their reproductive functions. There is an almost human sagacity in these actions; in the display of brilliant colors and soft perfumes to attract their lovers; in the cunning which they show in imprisoning a bee if he should arrive before the pollen grains are ready to be sifted on his back, and of holding him, sometimes days at a time, until he can go forth laden with the pollen that is to adhere to a pistil and so find its way to the ovary and perform the great miracle that results in seeds; in the many similar tricks which they use to entice and to hold; all working together towards that one great aim of plant life-reproduction.

"Perhaps one of the strangest and most interesting methods of securing cross-fertilization is that used by certain water plants which have their flower-stalks entirely hidden under the water. The Italian eel-grass (Vallisneria Spiralis) uses this unique method of fertilization. The female flowers grow on long, spirally twisted stalks, and each flower is enclosed in a small bladder. The male flowers grow in bunches, and each entire bunch is covered with a thin skin-like sheath. The female flower has continued to reach up her head until the flowers rest on the surface of the water, while the male is tied down below by a short stalk.

"Now the miracle happens! The gallant wooer deliberately breaks loose from his underwater position and arises to the top, where his lady-love is peacefully floating! The male flower bursts open his sepals and forms a tiny raft and, by means of this raft, he is enabled to float around until some kind of wind or wave brings him in contact with his love. Some of the pollen from the male adheres to the female flower; she drops to the bottom of the water, and there remains while the seeds are being developed."

"Plants, as this student of them affirms, after his careful investigation of the evidence, possess a psychic sense. There are numerous evidences of it in the plant's power to discover the presence of objects necessary to its welfare. A climbing plant, which needs a prop, will creep toward the nearest support. Should this be shifted to a spot several feet from its former position the vine will, within a few hours, change its course to the new direction. Is it possible that the plant sees the pole? Such a theory may explain the action in this instance, but if the plant grows between two mounds or ridges and behind the ridge stands a wall which will afford good climbing but is invisible from the position of the plant, while behind the other ridge is no form of support, the plant invariably will bend its course over the ridge which is before the wall. Examples of this may be found wherever climbing or creeping plants grow. The support is invisible from the plant's starting point. There is no odor which, as is possible in the location of water, might give the plant some clue to the direction in which its support may be found. The only explanation seems to be the existence in the plant of a psychic sense."

"There is at least one other sense which is possessed by plants in a marked degree. This may be called the physical sense. For example, most house plants which in their domestication have assumed more or less artificial forms, will, on being returned to their original haunts, reassume their original or natural forms. There must be in the plant some prompting sense which makes it realize any unfitness in its life or being.

"Plants, then, have seven senses: sight, hearing, feeling, taste, smell, a psychic sense and a physical sense; or six senses and a reasoning power—if the physical sense be admitted as such. These senses might be termed 'passive' mentality—that is, senses which, to perform their functions, possibly do not require any command of the will, but are merely natural to the plants. If, however, these seven senses are but passive powers, and not in any way an evidence of intelligence in the plant, there are certain actual and purposeful motions of the plant which might be called its 'active' mentality."

It is the purposive and intelligent acts of the plant that compel us to impute to them intellect, will and consciousness. When we consider the fact that the being that builds the plant is the same being that builds the animal, we should expect that either structure, plant or animal, should exhibit the same signs of being occupied by and guided by intelligence. The facts are that those in charge are intelligent, just as are those in charge of the submarine or the battleship.

Man is supposed to be the most intelligent, but compare his acts with those of the cell and you will be surprised at the number of foolish and useless actions performed by man.

Consider the actions of the plant called the Didinium, who lives a separate life in the water. The following by Mr. Binet is a description of his actions: "The most complicated instance of localization is met with in the Didinium, which we have so often cited; the Didinium knows precisely the position of the prey it follows, for it takes aim at the object of its pursuit like a marksman, and transpierces it with its nettle-like darts. Between these two species, we find all the intermediate instances of a localization of perceptions."

Now these are descriptions of the single cell, living the wild life, where he struggles for existence as a sole separate individual. Compare the actions of these hunting cells with those of Theodore Roosevelt, shooting deer and bear. There is not the slightest difference in the intelligence exhibited by the cell and by Mr. Roosevelt in their actions except in one thing, and that is, that the cell had to first make his own gun, so that in this particular, the cell has Mr. Roosevelt beaten. But you may say that a cell never does anything with the intention of effecting a purpose.

When we see an Indian make a bow and arrow, we know his purpose—to get something to eat. With that idea spurring him into action, he works for a purpose, When he shoots the arrow, it is for the same purpose—his every act is for the one purpose, to get something to eat.

The acts of the Amoeba and the Didinium must necessarily be the same. They are the actions of an intelligent being in either case. The cause of all life we see, is the intelligence possessed by the microscopic builders, the cells.

A short time ago I read in my Sunday paper the following article: "Botanists have long declared that plants as well as animals have nervous organizations and are capable of feeling and demonstrating that they feel pleasure and pain and even that they show appreciation for attention and droop under neglect.

Scientific proof of the truth of these theories is now furnished in records of remarkable experiments conducted by Professor Jagadis Chandra Bose, of Calcutta, India, with the aid of ingenious mechanism invented by himself. This instrument is called a "Resonant Recorder."

Writing about it in the Modern Review, of Calcutta, Professor Bose says: "There are rhythmic tissues in the plant, which, like those in the animal, go on throbbing ceaselessly. These spontaneous pulsations in one case, as in the other, are affected by various drugs in an identical manner. And in one case, as in the other, the tremor of excitation is transmitted with a definite and measured speed from point to point along fiber-like channels."

A writer in The Nation, London, describes the results obtained in England by the use of Professor Bose's recording device: "One of his delicate machines records the exact rhythm of the leaf's pulsations. A needle sets it down in dots on a piece of smoked glass. Then when the professor doses the prisoner with alcohol the curve becomes one of exhilaration. He gives it carbon dioxide and the plant grows ill, and signifies the same in its Morse code. He poisons it, and the pulse ticks dolefully lower and lower till it finally stops." I have to use these illustrations as arguments to overcome the natural objection and prejudice that the reader will have against believing that a microscopic being or a plant can possess feeling, pain and sorrow. It is hard for the reader to comprehend that the little cell is his builder and caretaker as well as that of the plant.

The following is an article from the Pittsburg Dispatch, which shows that some plants see, hear and speak:

"From far Brazil Harry J. Black, an American, is coming with one of the most remarkable collections of orchids, valued at more than $75,000, together with several hundred other rare plants. Mr. Black is a well-known editor-publisher of Buenos Aires, owner of Fraymocho, an illustrated weekly printed in Spanish. He is accompanied by his friend, Senor Diego Gibson, a native of Buenos Aires, likewise Senor Ramon Caceras, a horticulturist from Montevideo. They are bringing the Black collection of unusual plants for the purpose of exhibiting them at the third international flower show under the auspices of the Horticultural society of New York, Grand Central Palace, March 17-23.

"Mr. Black has spent twenty-seven years in South America. Flowers are merely a hobby with him, and he has made lengthy excursions thru the wilds of Bolivia, Peru and Brazil, emerging from the dense forests after a stay of several months, with plants never known to have been seen by human eyes. Mr. Black was one of the first to agree with Sir Francis Darwin when, as president of the British Society in 1908, Darwin delivered an address declaring that in plants there exists 'a faint copy of what we know as consciousness in ourselves.' The famous scientist was laughed at by many in his theory that plants can see and hear, but Mr. Black believed it and indulged in research work that would prove it, as did Mr. Jean Viaud-Brant, nurseryman of Poitiers, France, who maintained that the rose could see the beautiful woman inhaling its perfume, and furthermore that plants can hear.

"In proof of these theories, Mr. Black is bringing some specimens, one of which is a sensitive plant that folds up its leaves in fright if a sharp noise is made nearby, and the same plant is almost human in that in addition to having the temperament of a nervous woman, it is also rendered insensible by anesthetics such as ether, chloroform, heroin, etc. Its discoverer says that he has reason to believe that plants have a system of speech and, like Mr. Viaud-Brant, cites cases where the scent of flowers is a manifestation of their vegetable life and living radiation. Scent considered as an olfactory sensation is a vibration, and scent therefore is the sound uttered by flowers, or a song without words as it were.

"Among the astounding plants which Mr. Black will bring is a species of jatropha from Colombia, known as the ortiga brava albo (the cruel white metal) which secretes a poison like a rattlesnake, and when touched, two tiny organs which correspond to the tongue of a snake shoot poison and inflict a deadly wound. He also brings some varieties of stinging plants, which have long hairs, and when a hair is snapped by contact, it discharges poison in even sufficient quantity to kill a man. He intends to demonstrate this by killing mice and insects."

We see in every individual, plant or animal an effort to improve and maintain its stage of existence on this planet. Call it chance, or intelligence, or what you will, it is the same in plants and animals as it is in man, and we find it the same in the cell as in those individuals which he produces. However, we know that chance can not take care of a social community like a city or an animal. There must be loyalty, duty, sleepless watchfulness, a wisdom to make and unmake, and to keep careful watch over all that happens within and without.

Did you ever stop to consider all the wonderful schemes that plants have invented to cause their young or seeds to be scattered over the surface of the earth? It is such a common matter that we never stop to give it a thought.

The following is an extract from my common school Botany: "Dispersal by currents of air.—Many seeds are so light as to be carried about by currents of air. Ordinarily, however, the wind-dispersed seeds of fruits develop special appendages to aid in their flight, commonest among which are wings and tufts of hair. For example, wings are developed by the fruit of maples and elms, and by the seeds of catalpa and its allies. Plumes and tufts of hair are developed by the seed-like fruits of thistle, dandelion and many of their relatives and by the seeds of milkweeds, willow herbs, etc. On plains or level stretches, where winds are strong, a curious habit of seed-

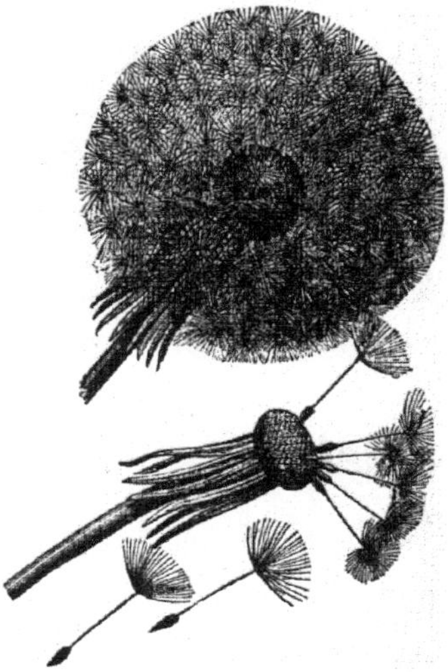

Fig. 34.—Seeds of dandelion with tufts of hair for parachutes.

dispersal has been developed by certain plants known as tumbleweeds or field rollers. These plants are profusely branching annuals with a small root system in light or sandy soil. When the work of the season is over and the absorbing rootlets have shriveled, the plant is easily broken from its roots by a gust of wind, and is trundled along the surface like a light wicker ball, the ripe seed vessels dropping their seeds by the way. In case of an obstruction, such as a fence, great masses of these tumbleweeds may be seen lodged against the windward side. This method of dispersal is far more effective than the mechanical discharge; but it is fitful, and its range usually is not very great. Thistle-down may be floated into a neighboring field, and a strong wind may carry the comparatively heavy-winged fruits of the maple and the elm some distance; but at best the scattering is only over

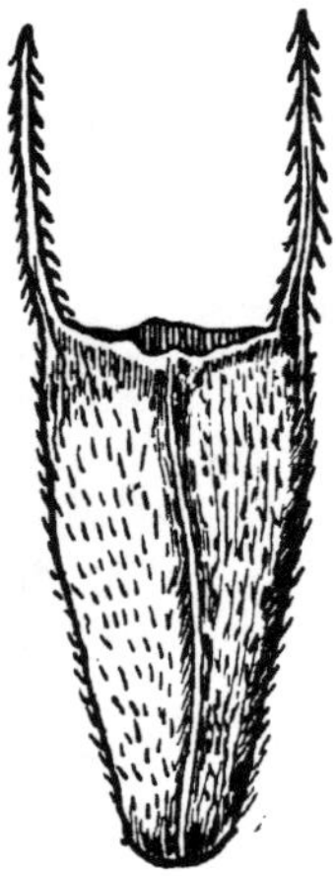

FIG. 35.—Seeds of beggar-ticks with barbed appendages.

a neighborhood. In many cases seeds or fruits or heads develop grappling appendages of various kinds, forming the various burs, which lay hold of animals brushing past; and so the seeds are dispersed."

The skillful performance of the Russian thistle and the tumbleweed is wonderful, while they are growing during the summer. You can not possibly pull a plant out by the roots, but as soon as it is ripe for seed dispersal, it is cut loose by the cells that made it, and stands ready to travel with the first good wind and is made into a round ball showing the intention and purpose of the builders. It is also so constructed that it hardly ever gets tangled and caught in grass and weeds.

The balloons or parachutes attached to the seeds of milkweeds and dandelion are certainly works of art. Is it not absurd to state that all these perfect designs and structures to effect certain purposes and ends could come about by chance? Why should it require any more intelligence to build a parachute by man than by the cell? One is a living animal just the same as the other, eating the same kind of food, breathing the same kind of air and made of the same kind of material. If the acts are purposive and intelligent in one case they must be in the other.

The following is a description of how the water plants are adapted for water life. This is how the cells build their house-like plants in water, in such a manner that they will be able to enjoy the comfort of both air and water:

"Adaptations.—When a plant lives entirely or partially submerged in water, its structure differs in many ways from that of an ordinary land plant, and these adjustments to water life are called adaptations. On parts under water the epidermis is thin and permits absorption, so that in a completely submerged plant its whole surface absorbs. When this is the case, the root-system is much reduced in extent as compared with a land-plant of the same size, for it is not the only organ for water absorption. In submerged plants the rigid tissues are less developed than in land plants, for the buoyant power of water helps to support the plant. This fact may be illustrated by taking from the water submerged plants that seem to be upright, with all their parts spread out;

upon removal they collapse, not being able to support themselves. Water-plants are also usually provided with air-chambers and passageways that the air may be free to reach the working cells."

In what manner do these structures differ from those of man under similar circumstances? Although the cell is an animal that lives naturally in water, still it also requires air and as it is evidently easier to extract oxygen from air than from water, it has provided air chambers filled with air under water. I have examined a number of water plants and every one of them has hollow open chambers under water which are always filled with air. Now some one must understand how to build these air tight chambers and fill them with air. The air chambers in bamboo rods are sometimes an inch across and five to ten inches long. Those who think that it is an easy matter to build and maintain air tight chambers under water and keep them filled with air had better try it and be convinced that the cell is just as skilled and smart in his line of business as any one else.

I do not pretend to know what intelligence is, nor what memory is, but I want to show that the cell is a being possessed of that something, whatever it is. If man is intelligent the cell must be. The cell is an active living animal, he mates, loves, feels, eats, drinks, breathes, jumps, moves, and performs all the things that every animal does,—that is, he has the essential attributes of a living being.

Before closing this chapter, I must describe a few of the hundred or more varieties of plants which have fly traps with which they catch and devour insects of all kinds. These plants should illustrate to the reader most clearly what a plant or animal really is. The reader should be able to comprehend from these plants provided with fly-traps that every plant or animal is simply a structure built for and occupied by smaller living animals and beings we call cells.

FIG. 36.—Leaves of the Californian pitcher-plant, showing the twisted and winged pitcher, the overarching hood with translucent spots, and the fish-tail appendage to the hood.—After KERNER.

FIG. 37.—Leaf of *Nepenthes*, showing the blade-like base, the tendril portion, and the terminal pitcher with its lid.—After GRAY.

The following is a correct description of these plants from text books on botany used in the high schools:

"A much larger California pitcher-plant is Darlingtonia, whose leaves are one and a half to three feet high, the hood bearing a gaudily colored 'fish-tail' appendage, the whole structure being a more elaborate insect trap than are the leaves of Sarracenia. In these traps not only are the remains of flies found, but bees, hornets, butterflies, beetles, grasshoppers and even snails have been reported. The species of Nepenthes from the oriental tropics, very common in conservatories, develop most remarkable leaves, the lowest part being an ordinary blade, beyond which is a well-developed tendril, at the end of which there arises an elaborate pitcher with a lid. There is the same sweetish secretion at the rim of the pitcher, and the same accumulation of water within as in the ordinary pitcher-plants. Leaves of sundews.—The sundews are forms of Drosera and grown in swampy regions, the leaves forming small rosettes upon the ground. In one form the blade is round, and the margin is beset by prominent bristle-like hairs, each with a globular gland at its tip. Shorter gland-bearing hairs are scattered also over the inner surface of the blade. All these glands excrete a clear sticky fluid, which hangs to them like dewdrops, and which, not being dissipated by sunlight, has suggested the name of sundew. If an insect becomes entangled in one of the sticky drops, the hair begins to curve inward, and presently presses its victim down upon the surface of the blade. In the case of a larger insect, several of the marginal hairs may join together in holding it, or the whole blade may become more or less rolled inward.

Leaves of Dionoea.—This is one of the most famous and remarkable of insect-trapping plants, being found only in certain sandy swamps near Wilmington, N. C. The leaf-blade is constructed so as to work like a steel trap, the two halves snapping together and the marginal bristles interlocking like the teeth of a trap. A few sensitive hairs, like feelers, are developed on the leaf surface; and when one of these is touched by a small flying or hovering insect, the trap snaps shut and the insect is caught. Only after digestion, which is a slow process, does the trap open again. Dionoea is popularly known as the 'Venus fly-trap.'

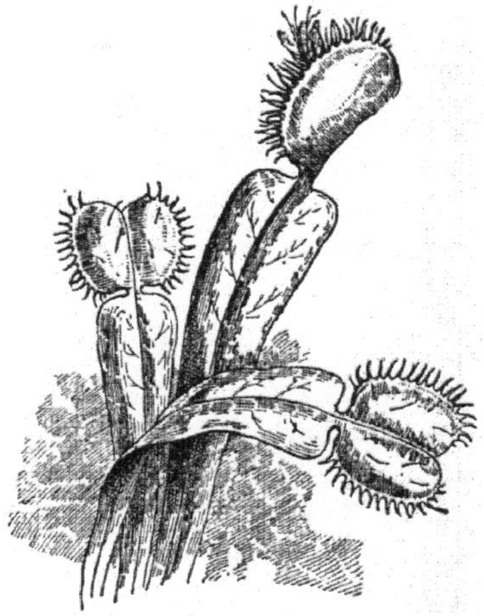

FIG. 38.—Three leaves of *Dionaea;* two with the traps open, one with trap shut on an insect.—After KERNER.

"Sarracenia, Drosera, and Dionoea are conspicuous representatives of the so-called carnivorous or insectivorous plants, all of which capture insects and use them for food. They are green plants, so that they manufacture carbohydrates; but for some reason they supplement their food manufacture with a supply of food already manufactured, and obtained from the bodies of captured insects."

I have examined all of these plants and several others. They stand as a living example of the inventive genius and intellectual capacity of the cells which occupy them. All of these plants understand how to make starch and food from the crude material of earth, air and water; there was no necessity for these traps in order to perpetuate their existence. They could not in any case have come to exist by slight beneficial variations of the leaves, because they would be of no benefit until completed as a fly-trap. The traps must have been fully planned in the minds of the builders before they took any steps toward their construction. Darwin's Theory of Evolution could have had nothing to do with producing this structure by beneficial chance variations.

In fact, in every case where I have investigated the theory of evolution's producing any structure, I have found that it does not produce. The theory of the survival of the fittest, in so far as its causing the existence of anything, is a myth. It is an incident that determines which of two shall live, but that is not necessarily the best man. It ignores the activities of the producer and does not pretend to point out who the producer is.

In a struggle for existence between two men, the stronger will not necessarily survive, the weaker is just as likely to survive. He may use poison, or some other unfair means against his opponent. The one with the most inventive skill and intelligence will generally survive. So in any way you examine it, intelligence is the cause of all living structures, whether they are houses, railroads or battleships built by man, or plants and animals built by cells. Someone must be there with the intelligence, as so far we have not found anything produced by magic or a miracle.

The singular and significant thing about these plants

with fly-traps is this, that it is not always the pure necessity of the case which prompts the cells to build any particular structure, as they can live without these insects, just as other plants do. They have produced these fly-traps for mere sport and luxury.

In this particular they also resemble man in his most highly civilized state. Where can you find any traps or snares invented and constructed by man to catch animals that can beat these for skilful construction and inventive genius?

Think of the trick employed to fool the insect in the California pitcher-plant. First the flaring colored flag hung out to attract his attention, to make him think it is a flower, then the sweet smell coming from the inside of the trap to tempt him to enter, the glazed surface on which he will slip, the spikes to keep him going straight on in, after he gets started, the transparent covering on the top, to fool him to fly or go in in the wrong direction, so that he never can get out. Then there is the water in the bottom where he will drown, and where the cells swim in and devour him, and feed his drowned carcass to the other cells in the body of the plant.

The cells who build this plant with this fly-trap attachment need take no back seat for any being on earth, even an inventor, Edison not excepted. The intelligence of the cell is the same in all places, whether he is in charge of a plant, insect or animal or living singly in the water and killing his fellow cells with weapons at a distance, or in the human brain directing the actions of the German army.

CHAPTER 7.

CAUSE OF HEREDITY.

It always has been and today still seems to be a standing mystery how the seed from a plant, tree, animal or insect can develop into the same kind of living structure as the one from which it came. Why does a kernel of corn develop into a corn plant and not into a sunflower? There is, however, nothing very mysterious about this when we consider all the facts in the case. The kernel of corn is a cell, a living animal or being. Where did he come from? What has been his former experience? What has he been doing and what does he know? When we look up his past history we find that he has been in the business of building corn plants for ages. He has been educated in that work and none other. He has been sent into the world with instructions to build the corn plant and has been provided with enough food and building material to get a good start in life. Mr. Ribot defines heredity as "that biological law by which all beings endowed with life tend to repeat themselves in their descendants." I think Spencer gives a better definition. He states that it is "the capacity of every plant or animal to produce other individuals of a like kind." The cause of heredity is the intellect of the cell based on his power of memory.

Every intelligent animal or being has a memory where

his past experiences are recorded, and with its help he will be able to repeat what he has experienced in the past. The animal or plant not only inherits the features, form and constitution of the parent but also the intellect; that is to say, it will also inherit the parents' habits, actions and mental qualities. The young muskrat or beaver will inherit not only the form of its parents but also their architectural skill. It will build houses over the water on the prairie marshes and streams, from mud and grass. The houses will be of such size as will be necessary to take care of the family. If it is a large family, a large house will be provided. The house of the muskrat will usually have two rooms; one will be a dry, cozy, warm place where they sleep, the other will be a place a little lower down where they eat. If not disturbed by man or mink during the winter they would lead a very cozy and comfortable life in the most severe prairie climate, in the fiercest snow storms, with a temperature forty below zero. In what way does this house of the muskrat differ from the log, brush and sod houses of man?

When the early settlers first came to Minnesota, they dug holes in the side of a hill and covered them with brush and grass, because they could not get any lumber. They were then cave-dwellers. The fox also dug holes in the ground, but he never covered his holes with anything. He was a cave-dweller also. The badger was more like the early settlers. He dug holes and covered them with earth and grass when he retired for the winter. Before any crops were raised, the early settler and the fox led a similar life. They both had to hunt for a living. Two or three years later, the early settlers improved upon this primitive cave-dwelling, and put up one on a plan similar to the muskrat. They built a two-room house on top of the ground out of the material at hand, which was earth, grass and brush. A little later on the railroad came through the country and brought lumber for better houses. This simply goes to show that man adapts himself to conditions, just as other animals do.

However, what I started to show was this, that the rat knows how to build houses over the water in which to live, and not houses in the trees like squirrels, nor houses in the ground like the badger. The building of the houses of the muskrats is an experience and ability possessed by this animal alone, that is, by the cells in his brain. They know how to build these and none other. The houses of the muskrat are always about the same. They are for a specific purpose. They are intended to provide the occupants with a place to live over the water in the severe winter, with entrances into the water under the ice, where they can obtain the roots, insects and grass in the water for food. They must be just so, or they would not answer the purpose. It requires considerable skill to build these houses because the thick ice, which will cover the marsh in the coming severe winter, must be correctly calculated. The singular thing about this animal, as well as other animals and birds, is that it knows how to build these houses without ever having seen one before, and without ever having had any previous instruction in the art. He is born with the knowledge. We shall see upon further investigation that this knowledge is located in certain cells of the body. The cells that build the rat not only know how to build him as a structure specifically adapted to live in water, but also know how to build another structure for his home, using him as a machine with which to do it. The young duck knows that it is proper to jump into the water, and goes in the first opportunity it gets, while a young chicken will keep away from the water. Without having had any previous experience or information in the matter they know from the beginning what is the right thing to do. Why? Because one is a structure made to move on both land and water, while the other, the chick, is made to move on land only. It would be absurd to think that the builders of these structures did not know for what purpose the structures were made.

The majority of scientists now seem to agree that there is no difference between instinct and intelligence. Mr. Ribot states, after having considered the different instinctive actions of man and animal: "There is therefore no absolute distinction between instinct and intelligence. There is not a single characteristic that remains the exclusive property of either." Then he cites as illustrations, among others, the actions of the bees, ants and wasps, and states: "Neither is instinct always so blind, so mechanical as is supposed, for at times it is at fault. The wasp that has faultily trimmed the leaf begins again. The bee only gives the hexagonal form to its cell after many attempts and alterations." Bees that have never seen the gathering of honey nor the building of the comb will go at the work as if they had been practising for years. A bird raised in captivity, who has never seen a nest before, will at the proper time build one just like its parent's if given a chance. These intelligent acts are called instinctive. This intelligence must be somewhere in the animal. The cell that caused the construction of the young bird had been in the nest building business for ages.

We do not need to know how life originated nor what it is in order to know the cause of heredity and of development of life, because we can see by the power of the microscope that certain animals or beings we call cells are the builders of all those living things that we can see without the microscope. In the same manner we do not need to know who built man in order to know who built railroads, skyscrapers and ships. It is not material to the question of the cause of heredity and development, to know who built the cell.

If we find that this microscopic animal is the builder of all these things, then the only question is, why is it that one cell will produce this structure, and another cell, which looks just like him, will produce one entirely different? When we consider that he is a living animal just like ourselves, and just like the thinkers in our head, we need only ask, why do some men build sod houses and some skyscrapers, and why does a squirrel build in trees, and muskrats over water? The answer is plain—one knows how to build one kind and the other, another kind. It is simply a matter of education and experience of the cells. Mr. Walker states:

"Every part that is alive of an animal or plant consists of cells and of nothing else. There are parts of the bodies of animals and plants that are not composed of cells, but these are not alive. The hard parts of the bones in man are an example of lifeless substance within the body. Such dead material, however, has been produced by cells, which form a part of the animal or plant during its life."

Investigations have shown that a person will inherit the same shaped bones or frame-work as his parents. Bones are not alive—they are only the structure built by the cells from lime and other material to serve as the frame or supporting structure. The cells station themselves here and there in the bone with the purpose of keeping them in repair. The following by Chas. Walker will start us in the investigation of the cell a little more in detail: "All the multicellular organisms commence their existence as single cells. This single cell divides

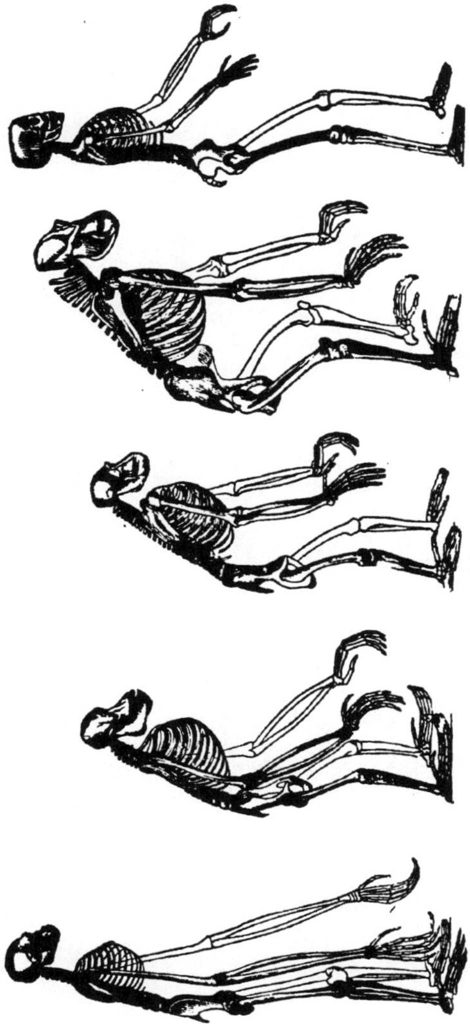

Gibbon. Orang-outang. Chimpanzee. Gorilla. Man

FIG. 39.—Skeletons of man and apes.—HAECKEL.

into two cells. Each of the two thus produced divides again, and this process continues until the whole body of the multicellular animal or plant is built up. Among the cells of the earlier generation there is a great similarity

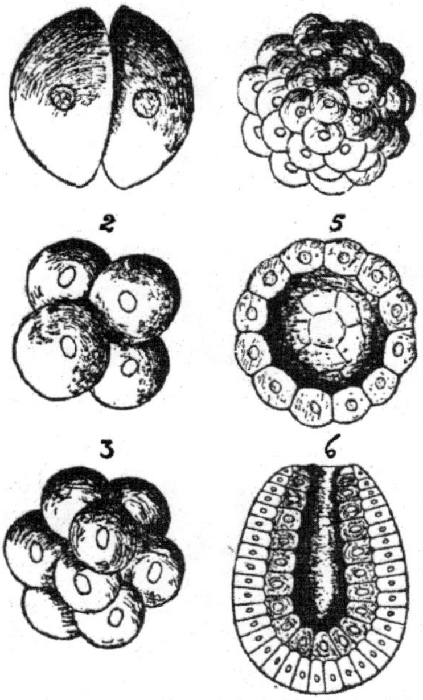

FIG. 40.—How cells multiply and begin building an animal: 4, morula: 5, section through blastula showing hollow sphere; 6, gastrula showing outer layer of cells (epiblast) and inner layer (hypoblast); the 6 is at the mouth of the cavity (enteron) of the gastrula.—HAECKEL.

in appearance, in fact, until a great number of cells have been produced, it is impossible to see any difference between them. As in the higher animal, the cells that are eventually going to be thrown off to form new individuals, the sexual cells, are differentiated at a very early period, frequently long before birth. These cells live a parasitic existence in the body, which has produced them, and in a certain sense can hardly be regarded as forming an integral part of it. They are certainly not necessary to the individual, but they are necessary for the production of new individuals and are thus essential to the race.

"The body of the multicellular organism is practically a colony of unicellular organisms living together and dependent upon each other; and with certain limitations this idea is sufficiently near the truth to be very helpful in obtaining a proper idea of the nature of a multicellular organism and of the transmission of character from parents to offspring.

"We may for the sake of clearness divide the whole of the living organisms into the two great groups—the unicellular, those in which an individual contains but a single cell, and the multicellular—those in which the individual contains more than one cell. The vast majority of the unicellular organisms are invisible to the naked eye, but nevertheless they perform the same function of digestion, secretion, excretion, etc., as the whole body of a multicellular organism, which may be built up of many millions of cells."

You notice the fact that even before birth a certain group of cells is set apart and a separate place is provided for them in the body where they live only a parasitic life. That is to say, they do absolutely nothing but study the subject of how to make the next body and how to improve it, if possible. These cells begin their training and education the first thing, and occupy their mind with nothing else. They are in continual mental touch with every part of the body, and can send out messengers at any time for information that they may want; or they can make personal excursions themselves by riding in the blood stream at any time or to any point or part of the body, and take note of this or that, just as the other white cells do now, who are the soldiers and general inspectors of the body. When we consider what the cell really is, that he is in fact a colony of beings, that half of him has the experience of ages and the other half of his crowd has the experience of the last body he occupied he should be very well informed indeed. You see, however, that this knowledge must necessarily be limited to those bodies or structures from which he came. The world in which he lived was the body of the plant or animal from which he came.

We might investigate here a little further into the details of how the cell multiplies and grows. We do not know how he grows but we know how he multiplies. He simply divides himself into two parts, then these two halves again grow back to full size, and then these again divide into two and so on. Figure 41 illustrates the way he looks through the most powerful microscope now made. He has a great number of special organs, the purpose of which we do not yet understand. We do know that he has a centrosome acting like a general superintendent, which looks after the division of the individual cell and sees to it that the division is exactly equal, as far as the central part of the body or head is concerned. This central head is called the nucleus, and seems to be made up of a crowd of separate individuals called granules, which are no doubt the living primordial beings that make up the cell because they move about in obedience to the orders of the centrosome. The centrosome seems to be the main head or directing center, the general manager, and the nucleus seems to contain the sub-managers, or skilled workers, and the body or cytoplasm, as it is called, consists of the common laborers or unskilled workers of the body of the cell.

The general appearance of the cell would indicate him to be a highly developed and specialized being made up of multitudes of smaller primordial cells or beings. They

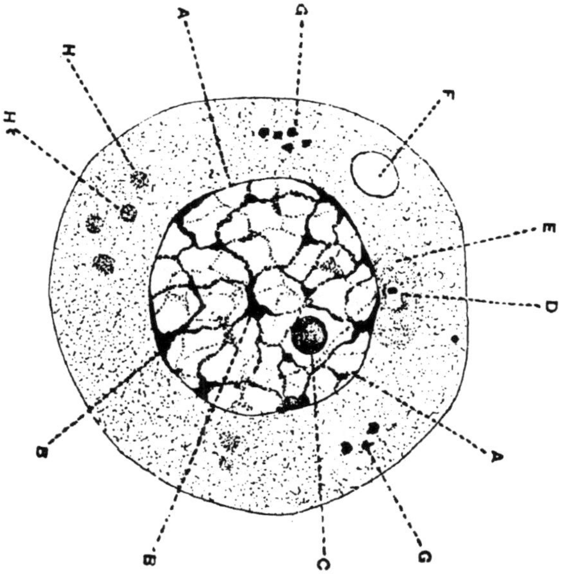

FIG. 41.—The cell. *A*, nuclear membrane. *B*, masses of chromatin, joined by threads of linin containing chromatin. *C*, Nucleolus. *D*, Centrosomes contained in *E*, the archoplasm. *F*, Contractile vacuole. *G*, Food particles. *H*, Plastids.—WALKER.

seem to co-operate and work together in a social way like the cells of our body. The wonderful thing that we observe is this, that when the centrosome gives orders to divide, they line up in rows like soldiers and the crowds divide exactly in two. There are evidently two or more beings or possibly crowds of beings in each granule, so

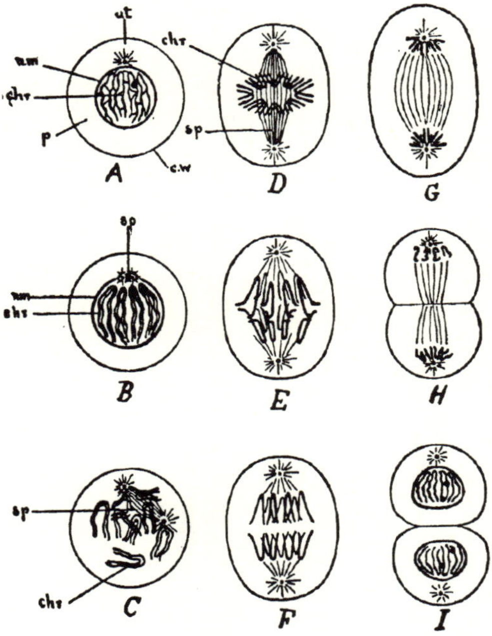

FIG. 42.—Diagram illustrating Mitosis or cell division. *A*, the cell commencing activity; *B*, *C*, *D*, phases in the formation of the spindle and the chromatin loops or V's, also showing that the mother V's have split into daughter V's; *D*, the chromatin loops forming the equatorial plate, *chr*; *E*, *F*, *G*, separation of the daughter loops (daughter chromosomes) and their passage towards the poles of the spindle, thus forming daughter nuclei; *H*, *I*, division of the protoplasm so as to form two daughter cells; *at*, attraction sphere enclosing a centrosome; *n m*, nuclear membrane; *chr.*, chromatin threads; *p*, protoplasm; *c w*, cell wall; *sp*, spindle.—SCHUTE.

that when the cell divides each half gets its full number of individuals, skilled and otherwise.

I quote from Walker: "What has been said with regard to the selective mode of division, which insures that an exactly representative half of each chromosome is handed on at each division applies to the linin even more forcibly than to the chromatin. The mode of division in the chromosomes seems to insure that half of every individual portion, of every individual chromosome, will be handed on throughout the succeeding generation of cells, for the division of the chromosome does not appear to be merely a nonselective division of the bulk of substances forming it."

In regard to the same point Wilson makes the following remarks: "The splitting of the chromosomes is therefore in Boveries' words, 'an independent vital manifestation, an act of reproduction on the part of the chromosomes, an independent reproductive act of the chromatin. The construction of the nucleus and in particular the breaking up of the chromosome after division into small granules, and their uniform distribution through the nuclear cavity, is in the first place for the purpose of allowing a uniform growth to take place and in the second place after the granules have grown to their normal size, to admit of their precisely equal quantitative and qualitative division.' These observations certainly lend strong support to the view that the chromatin is to be regarded as a morphological aggregate—as a congerie or colony of self propagating elementary organisms capable of assimilation and growth and division.

"Summary in conclusion—All cells arise by division from preexisting cells,—cell body from cell body, nucleus from nucleus, plastids from plastids, and centrosomes from centrosomes. The law of genetic continuity thus applies not merely to the cell considered as a whole, but also to some of its structural constituents."

On this point there seems to be no question or disagreement. The following by Mr. Wilson is also interesting, showing his view of how the cell, by equal division, is able to pass on continually from cell to cell every experience and tradition of the ages. A record is taken of the past and preserved, and continually handed on to the next generation. The following by Prof. Wilson will further illustrate this and also show that the centrosome is the intelligent directing center of the cell, and that if the nucleus is disconnected no work will take place. He says: "The interpretation of cleavage as a process of cell division was followed by the demonstration that cell division does not begin with cleavage, but can be traced back into the foregoing generation, for the egg cell as well as the sperm cell arises by the division of the cell pre-existing in the parent body. It is therefore derived by direct descent from an egg cell of the foregoing generation and so on ad infinitum.

"Extending backward from existing plant and animals to that remote and unknown period when vital organization assumed its present form, life is a continuous stream. The individual body dies, it is true, but the germ cells live on, carrying with them, as it were, the traditions of the race from which they spring and handing them on to their descendants."

A peculiar thing takes place when the cell divides. The granules in the nucleus line up in groups or strings which are called chromosomes, which we see from the following by Mr. Wilson are always of the same number in every plant or animal or species of cell. He states: "The remarkable fact has now been established with high probability that every species of plant or animal has a fixed and characteristic number of chromosomes, which regularly recur in the division of all of its cells,—and in all forms arising by sexual reproduction the number is even. Thus in some of the sharks the number is 36, in certain gastropods, it is 32; in the mouse, the salamander, the trout and the lily, 24; in the ox, guinea pig and in man, the number is said to be 16, and the same number is characteristic of the onion." It is very likely that these chromosomes represent different departments of the specialized individuals.

A conclusive proof that it is no chemical proposition but that the cell is an active, intelligent being and knows what he is about to do, is the fact that if you destroy the original cell or any one of them up to the number of fifteen cells after the first division, any one of the remaining cells will go on and build the body and finish the work just the same. This experiment seems to me to be conclusive proof that the cells build and work entirely from memory. They know what they are there for and what they started out to do, and whether one or more are accidentally killed in the beginning does not stop the rest of them from going on with the work and completing the structure. The following by Mr. Walker also tends to prove the proposition:

"On the other hand, Roux also found that in his experiments, when carried on further, the existing half embryo restored more or less completely the missing half. Later experiments by other observers were made with the eggs of several other animals, which appear to show that in the earlier stages of development, at any rate, all the cells into which the fertilized ovum divides retain the power of producing all the tissues that would under ordinary circumstances be produced by the fertilized ovum itself. Driesch, Morgan, Wilson, Zoja and others have separated the cells produced by the division of the fertilized ovum when development had gone as far in some cases as the sixteen cell stage. These experiments seem to prove that the characters cannot be represented by entities that are distributed in a selected manner among different cells during the process of development, as assumed by the Roux-Weisman theory. If it were so, this selection of different entities must begin at the first cell division, but it has been proved by experiment that even when the sixteen cell stage has been reached, each of the sixteen cells possesses within itself the power of producing, not only the tissues, which it would produce under normal conditions were the ovum left to itself to develop, but when separated from its fellows, also all of those tissues that would have been produced under normal conditions by the other fifteen cells."

This also shows that in division each one is able to pass on to the next one in some way the record of past events and experiences, which evidently is what we call the power of memory. In what manner the record is taken and kept we do not know. We can clearly see that they are very strict and careful in seeing to it in division that each one gets his equal half of the entire colony of primordial beings which make up the cell. It is evident also that the force in the cell is kept in duplicate. We do not understand the phenomenon of memory nor is it necessary in order to understand the cause of heredity and development. We know memory to be a fact, because the cells in our head can remember what took place in our childhood fifty or eighty years back. We know that there cannot be such a thing as intelligence without memory. We know that all animals, including man, reason and act from past experiences.

The cell is an animal. From the facts based upon what we see the cell do in the act of division, we are forced to the conclusion that every cell is able to remember not only what he has experienced during his own life but also what took place in the immediate generation before him. Another experiment also goes to show that it is not merely a chemical action or blind force which guides and directs the cell in the construction of plant or animal, to-wit: After the cell has started to build a plant or animal and it has multiplied into a bunch of say 15 to 100 individuals and they have arranged themselves in their proper places and begun the building of the different parts, if you then mix them up and flatten them out, disorganize and dislocate them, they will again scurry around and find their places and go on with the work as soon as possible. In reference to this Wilson states: "To both these tendencies is related the growth-process to which the future embryo will owe its form and every attempt to explain the position of the cells and the direction of cleavage must reckon with the morphological process taken as a whole—not merely a cell, dividing under the stress of rude mechanical conditions; it is beyond this, a builder who lays one stone here, another there, each of which is placed with reference to future development. Yet such eggs when released of pressure continue to segment without re-arrangement of the nuclei and give rise to perfect, normal larvae."

Driesh and Hartwig say, "The cells produced by cleavage are completely equivalent and indifferent and they may be thrown about like balls in a pile without the least degree impairing thereby the normal power of development."

It seems to me this should settle the question as to whether the actions of the cell are guided by intelligence or whether by merely blind chemical force. If you tear down an ant hill it will again be rebuilt. The same will be done if a wind or some other cause tear down structures produced and occupied by man. Structures produced or torn down by the blind forces will not be rebuilt. Notice that the author stated that the builder, the cell, is a person who "lays one stone here, another brick there, with reference to future development." How does a skilled architect proceed to lay the foundations for his structures? He would also do the same thing. He would lay one stone here, another there, having in view and in mind all the time what kind of a structure it is going to be when it is finished. It is just as necessary for the cell in building the skeleton on which the rest of the body will be supported that he place and arrange the material and particles of lime just so and all in the right place, keeping in view what he intends to ultimately accomplish, as it is for the skilled architect in building a house or a machine. The cell must build the skeleton from lime, phosphorous, etc., the ingredients must be mixed in proper proportions and placed in the right place. So must man mix the materials of concrete and stone and place them in the right place. The builder must have a mind that knows what is necessary and required at every step. The following by Mr. Wilson will illustrate how the cell in building animals or plants must have the material with which to build or else it cannot produce another individual like the one from which it came. He says:

"Every little organism at every stage of its existence reacts to its environment by physiological and morphological changes. The developing embryo like the adult is a moving equilibrium—a product of the response of the inherited organization to the external stimuli working upon it. If the stimuli be altered, development is altered. This is beautifully shown by the experiments of Herbst and others on the development of sea urchins. Pouchet and Chabry showed that if the embryos of these animals be made to develop in sea water containing no lime salt, the larvae fails to develop not only its calcareous skeleton but also its ciliated arms and a larvae thus results that resembles in some particulars an entirely different specific form." You will notice that if no lime is at hand the cells cannot produce the skeleton.

Before going into further discussion of development in general, I wish to call the reader's attention again to the fact that the cell is an animal divided into three general departments, the centrosome, which is the general superintendent, the nucleus, which seems to represent the skilled workers, and the cytoplasm or the main body, which does the general labor such as muscular work, etc. This shows clearly that the animal has special organs and that there is division of labor. Mr. Wilson calls our attention to it in the following language after fully discussing the matter: "The facts reviewed in the foregoing pages converge to the conclusion that the differentiation of the cell-substance into nucleus and cytoplasm is the expression of a fundamental, physiological division of labor in the cell * * *." Hemmingway concludes that "the centrosome is the motor center of the kinoplasm, both for the external and for internal manifestation. Lenhossek regards them as motors for the control of ciliatory action as well as for the spermatozoon and perhaps also for that of the nucleus fibrillae." Zimmerman concludes that the micro-centrosome is the motor center of the cell, also that it controls ciliary action. It is important to have a general idea of the make up of the cell and to remember that he is a very highly organized and specialized being. First there is a general manager called the centrosome, next a multitude of skilled workers or submanagers called nucleus or chromatin granules, and next, the general laborers or workers called the cytoplasm.

I wish now to call the reader's attention to one very important discovery and that is in reference to the future actions of the cells in the building of the animal. Why do some cells know how to do this work in the body and some another kind of work? I stated before that the nucleus contained the crowds of skilled workers, who are informed and trained in all the work necessary to build the entire animal. There is evidently in the nucleus of every germ cell a multitude of primordial cells with specific training for each specific department of work to be done in the body, such as brain work, digestive work, muscle work, respiratory work, etc. When the cell undertakes to perform some specific work, for instance a brain cell or liver cell, he has no use for those other skilled workers in the nucleus that are not necessary to help him in his department, and so they are excused from further service.

This was a very important discovery, and Mr. Wilson has the following to say in this matter: "Boveries' remarkable observation on the nuclei of primordial germ cells demonstrates the truth of this view in a particular case, for here all the somatic nuclei lose a portion of their chromatin and only the progenitors of the germ nuclei retain the entire ancestral heritage. Boveries himself has in a measure pointed out the significance of his discovery, insisting that the specific development of the tissue cells is conditioned by specific changes in the chromatin that they receive. It hardly seems possible to doubt that the limitation of the somatic cell in respect to the power of development arises through a loss of particular portions of the chromatin. Its application to development becomes clearer when we consider the nature of the 'nuclear control' of the cell, i.e. the action of the nuclei upon the cytoplasm."

This goes further to prove what a highly organized and specialized individual the cell really is. Just consider its actions! In the first place, each new one takes his place as it multiplies. One group stations itself where the leg is going to be; others where the brain and head are going to be; another where the liver will be, and so on Each group goes to work in harmony and with an understanding with all the others.

Let us now briefly consider what problems the cell has to contend with, in the building of an animal or man. It must remember the correct place to begin and the position of every part must be calculated to the minutest fraction in its relation to other parts. We see from their actions that the true position and place of every part must be profoundly inscribed in their memory. Even if some accident happened to displace them, they do not get mixed up and forget the true position of the parts, but they take their correct positions again. Perfect plans and outlines of the structures must be indelibly fixed in their memory. In this little room we call the womb, provided specifically for their work, sheltered from the elements and disturbances of the outside world, they must lay the foundations of the future structure, keeping in mind that their structure must soon be moved outside. There must be assurance of memory, accurate calculation, skill and faithful industry. This multitude of independent intellects accept their work and proceed in perfect harmony and complete their allotted tasks, if possible. They never get discouraged, disappointed nor lose faith in the future or the purpose for which they start to work. In a room with nothing to direct them, they must lay the plans for the future cell republic, and correctly they must mark out the locations of the parts of the machine or habitation that must be produced as quickly and economically as possible. In this labyrinth of complicated parts existing only in their memory, the numerous requirements of the large cell colony must be considered. They must remember how the different needs of the cell colony or city were taken care of in the body from which they came. They must remember how the special committees were appointed to take charge of this and that department. They must remember all the different things necessary to run the social organization of the cell colony which we call animal. The gathering and storage of food, the distribution of the same, general information, policemen, tradesmen, etc,—all this work must be taken care of and pushed along according to the records of the past. The streets, passages and stores must be placed at such places, and made of such material as they think best, that is in accordance with what they can remember from their past experience. How do we know that the cell has the power of memory? We can prove it in many ways. First, we know that the cells in our brain—our thinking cells, have the power of memory. We know that a brain cell is no different from any other cell of our body. Ideas come and go in our minds. My acts of yesterday or twenty or fifty years ago appear before me and go away like actors on the stage or in a moving picture. When the actions took place fifty years ago, they were preserved in a record in such a manner that they are ready and available for future use and reference. We speak of the unconscious life or unconscious mind, and unconscious memory. It would be more correct to speak of the mind and memory of those cells of the body that are not immediately connected with our senses. It is now, however, generally understood that memory belongs to both consciousness and unconsciousness.

The muscles of the body could not be trained to perform any difficult act if they did not possess the power of memory. I begin tc run my automobile at first very slowly—every act must be directed with my consciousness. After a while certain nerve centers or cells learn what is wanted and take full charge of the business of running the auto, so that my consciousness (which is the cells in charge of my sense organs) can attend to something else. So it is with walking, balancing or playing a piano. These acts are taken charge of by certain cells or little brains, as they are sometimes called, and it is the consciousness of these cells that has charge of and directs the various actions of the muscles of the arm in steering the auto or playing the piano. All movements we perform are the results of long difficult practice, except some inborn actions we call instinct.

The medical profession now well knows the reason why a person can become immune to any particular disease. The immunity is based on the memory of the cells of our bodies, especially the white cells or those whose business it is to look after the welfare of the body in general. They can remember for fifty years any experience in the past in fighting such dangerous enemies as the germs of typhoid, diphtheria, smallpox, etc., and whenever any such bacteria get into the body after having had a previous experience with them they do not allow them any chance to live, multiply and spread into the body, but they are immediately attacked and exterminated by these white cells as fast as they appear. Before they have had an attack by disease germs, and suffered the serious and dangerous experiences resulting from such attacks, they are generally careless, so that the disease germs get time to multiply and spread all through the body before their dangerous character is discovered. Relying upon these facts the medical profession make "serums" which are merely dead bodies of disease germs suspended in a liquid and when injected into the blood stream they frighten the white cells by leading them to think the disease germs have gotten into the body at some place, and they immediately prepare to fight the disease. However, we find that no "reaction," as they call it, takes place unless the body cells have had some previous serious experience with the particular germ injected. For instance, if a person has been sick with tuberculosis, and never had typhoid, the injection of dead typhoid germs will create no disturbance in the body but the dead or live tubercular germs will do so, even fifty years after the disease was cured. They remember their experiences fifty years back.

I can remember how my consciousness had to direct each finger when I first started to play the piano. Now consciousness needs only give a hint as to what is wanted and my nerve centers take full charge of the matter and execute the necessary acts. If my consciousness should be compelled to direct every detailed muscular act or motion, nothing could be accomplished worth while. If the judge in the district court had to look after the execution of all his orders and judgments, he would have no time to do the judging and directing. If the cells of the body did not all possess the power of memory, how could they be instructed and trained to perform the different acts required.

The mind of any individual is entirely based on his memory. All our ideas and conceptions are based upon it, we could form no ideas or judgments without a past experience to refer to. We know from the way the cells are connected and in touch with each other, that what one knows can easily be known by the others, if it should be necessary to their business. We notice how they work together in the case of sex instinct. The nerve centers in charge of the business of perpetuating the race are able to make the thinking cells connected with one's senses or consciousness believe that a certain female is absolutely indispensable to one's existence. The sex cells are able to present the female to the thinking cells as the only thing worth living for. The nerve centers directing the muscular movement of your arm steering the automobile must have a mental picture of the ground ahead at all times just as your conscious mind must have. The picture of the ground is transmitted direct to the nerve cells in charge of the special business of guiding the automobile. In the same manner the picture of the surroundings is transmitted direct to the skin cells of the fish that is able to change his color at will. This shows clearly how the cells are able to keep each other informed in every line of action.

As bearing upon heredity, it is interesting to note that the cells in the sex organs set apart to build another being at the proper time are in very close communication with every part of the body. They are in close touch with each other. Every cell has the same power of memory as the other. We know mind cells can remember acts performed by the individual in his childhood. The cells build from memory. That is also proven by the fact that no living being or animal is able to remember just exactly, but he will remember well enough so that his productions will resemble his previous productions or the place he came from more closely than any other productions. He cannot produce a structure just exactly like the place he came from, but it will be as near to it as he can remember. This fact will apply to all living beings from the cell to man. You can generally tell by the actions of a person and from his business what he is, what his experiences have been and what is recorded in his memory. When we remember the manner in which the cell multiplies we can see that his past experiences are passed on from generation to generation like a continuous record. It seems to me very clear how one individual can produce another one like himself when we consider the facts. Every organ in the body as a whole can be trained; the hand can be trained to do this and that until it remembers just how to do it. For instance in writing, every cell of the hand must remember the different movements. You can train the eye, arm or leg by repetition, just as you can the mind or thinking cells. One will learn as well as the other. Memory is a fact. It is a faculty possessed by all the cells. We know by our own experience that our mind is a record of the past. The records of past events and experiences are piling up all the time. The records are there, but the most difficult thing to do sometimes is to find the one we want, and we say we have forgotten.

It has been proven that cells are always of the same size. In reference to this Wilson states: "Measurement of the cells from the epidermis, kidney, liver, then alimentary epithelium and other tissues, shows that they are on the whole as large in dwarfs as in giants, and the same appears to be the case in the plants." In the same manner the size of a house or an ant hill has no effect on the size of the individuals that build and occupy it.

Walker states that it is easy to see how individual characters can be transmitted from parent to offspring. He states: "How this happens in the case of the individual character is easy to see, for we have individual entities, the chromosomes, that are distributed from cell generation to cell generation, from parents to offspring, in a manner that coincides exactly with the behavior of the individual character."

We come now to consider the question of the transmission of individual characters. There are some who claim that the action of the environment will cause variation and different characters in the offspring. This proposition is true to a certain extent, that is, if it is a cold climate the length of the hair will be long; if it is warm, it will be short. In the same way man will vary his houses and his bodily covering as to best serve his needs, but to say that it is the climate that puts up the house or makes the coat and puts it on the man's back is simply absurd. It is the intelligence and industry of man that makes the coat and builds the house. In precisely the same way, if it is necessary, the cell communities are able to change or vary the structure. Regarding this matter Walker states: "The action of the environment upon an organism has been claimed by some writers as the cause of variation in its offspring. In considering this question, it is necessary to have a very clear idea of what it means. We have seen that the environment may produce very great modifications in the individual. These modifications are acquired characters and appear in the individual at different periods of its life in response to stimuli from without. They are not inborn characters and unless the necessary stimulus is applied they will not appear." You notice his statement that if these modifications are not necessary, they do not appear. In the case of man, if he does not use the shovel and the hoe, but only the pen, the inside of his hand will be nearly as soft as any other part of his body. However, if he should change his occupation, the inside of his hand will develop a thick horny covering. Why? Because if the skin cells inside of his hand did not build up these buffers, the spade handle would very soon cut through the skin and tendons and destroy the hand. The evolutionist will say that the work and the spade handle provided for and built up those callouses in the man's hand. The climate will not build hair or make a coat nor will the spade handle build a horny covering to protect the inside of the man's hand.

When I helped load a wagon with concrete tile, I had to stop and get a pair of gloves, because the rubbing of the rough tile against the end of my fingers soon wore away the protecting outer skin, and unless I had provided protection in some way, the tips of my fingers would soon have started to bleed. The callous under your feet and the sole leather in your shoes are produced for one and the same purpose. The ground will not produce shoes or callouses to cover your feet nor will the tile produce gloves or the callous covering to protect your hands. Man must produce the shoes and gloves and the cell must produce the callous.

We might again consider some of the sexual characters referred to heretofore. Mr. Walker has the following to say on that subject: "Secondary sexual characters are those which though appearing in all individuals of the same sex and not in those of the other sex are not connected directly with the sexual function, that is, with reproduction. Such characters are the beard and voice of a man and the antlers of the stag. There is a great deal of evidence which suggests that the potentiality of producing these characters is present in the individuals of both sexes. Among mammals and birds we constantly find that the characters of the young male are those of the female minus of course a few special characters. The plumage of the young cock pheasant is similar to that of the adult female. A boy's voice is similar to that of a woman. If during infancy the sexual glands are removed from a male animal the male secondary sexual characters do not appear. Not only are the physical characters such as changing the voice and plumage inhibited, but mental character such as pugnacity do not develop. There are some direct experiments which suggest very strongly that the appearance of the secondary sexual characters is dependent on the presence in the body of the sexual glands."

You will notice that it is the same also with these sex characters. They are not produced unless necessary. The history of the human race shows that man battled with his fellow man for the possession of the female and in this battle it was an advantage to employ these masculine characters. It is also plain that it was a method of preparedness, both offensive and defensive, instigated by the sexual cells as soon as they felt ready to possess or defend a female. We find that if the instigators are removed no sexual characters are developed.

In insects we have the most wonderful illustration of the tearing down or destruction of one structure and with the same material rebuilding it into a new and different structure. For instance, the cells first build a caterpillar or worm; this worm as a structure is demolished by those same cells who put it up and a new and different structure is put up with the same material, known as a moth or butterfly. In some other cases, if the worm is cut in two, each half will be torn down by its occupants and the half will again be rebuilt into a complete worm, but each worm will be only half the size by reason of the lack of material. There is nothing mysterious about this because the structure is of no use as it is, so the only thing to do is to tear it down entirely and rebuild it. Man would do the same thing if someone cut away half of our house and we had no other material to fix it with, the only thing to do would be to tear it down entirely and with the material build a smaller one. Wilson has the following to say on this subject:

"Morgan's remarkable observation on 'planaria' finally shows that here also when the animal is cut into pieces, complete animals are produced from these pieces but only in small degrees through the formation of new tissues and mainly by direct remolding of the old material into a new body, having the correct proportion of the species."

Coming back to the question of variation I quote the following from Mr. Walker: "Variability appears to be a property common to all living organisms in spite of the fact that individual animals and plants produce new individuals that are generally similar to themselves. For instance, if a collie dog be mated with another collie, the pups produced will grow to about the same size as their parent, they will have similarly shaped heads and be similar in general. Every individual in the litter of pups will differ in some way from its brothers and sisters and also from its parents. But though these differences are very evident, upon careful examination they are in the overwhelming majority of instances comparatively small differences. The pups will in fact, though differing from their parents, still resemble them beyond all comparison more nearly than they will resemble a fox-terrior or any other breed of dog." Now this is just what we should expect, that the cells would build the pups as nearly like the parents as they could remember, but we know that in building anything from memory, it is not likely that the structures will be exactly alike. However here is a different proposition stated by Mr. Walker: "The crystalline lens of the eye is produced from epiblast but when it is removed in the salamander the new lens grows from cells that were produced from the mesoblast. Under normal conditions mesoblast cells in the salamander would never become modified into anything at all like the crystalline lens. Here the mesoblastic cells have still retained the general potentiality of the fertilized ovum in a very high degree and are able to reproduce such a specialized structure as the lens in spite of the fact that under normal conditions this is only produced from epiblastic cells.

"Some butterflies vary greatly in their appearance at different times of the year, so that they are protected by their similarity to their surroundings in both the wet and dry seasons, when conditions vary enormously. The difference is so great in some cases, that the wet and dry season phases have been classed as different species. 'Naturalists were fairly astounded when in 1898 Mr. Guy K. Marshall first bred the black and blue dry season Preus sesamus from the black and red Precis natalensis. The two butterflies differ in size, form, pattern, colors, relation of upper to under surface and habits."

This again certainly illustrates the same fact that the cells will produce such structures and habitations as will be necessary to protect their lives or that may be required in any particular place or under particular circumstances. Like does not produce like, as is the case in chemical and natural forces, but such things will be produced as will be required and are reasonably necessary. Such color will be provided as will most likely deceive his enemies and save his life and the same will be done by a general in the English army. The mesoblast cells understood how to make a lens for the eye, and when it became necessary it was produced.

The following by Mr. Walker in reference to the actions of some insects is also interesting: "There is a phenomenon in nature so striking and so general that no theory of evolution can be accepted as plausible, or even possible, which fails to explain it. This is the extraordinary adaptation of living organisms to their surroundings. Not only are they adapted to their physical condition but above all to each other. The mutation hypothesis not only fails to explain this, but a careful consideration of its postulates shows that the co-adaptation of living organisms to each other and to their environment and the origin of species by mutation are incompatible. There is no living organism which would not serve as an example of the phenomenon of adaptation. We must, however, consider a few individual cases in order to realize its full significance.

"Sitaris humeralis, a beetle belonging to the family cantharidae is a parasite upon the solitary bee, anthophora. The female Sitaris lays over 2,000 eggs, burying them in the earth near the entrance to the nests of the bee. These eggs hatch, producing larvae, which possess six legs, as is usual in the larvae of beetles. The larvae are triungulins, that is, they possess three claws at the extremity of each leg. This is exceptional among beetle larvae. The larvae hibernate until the following spring when they become active. They do not, however, try to enter the nests of the bees, but attach themselves to any hairy object that happens to approach them. No discrimination is shown in the choice of an object beyond the fact that it must be hairy. The majority of the larvae are doomed to extermination for they attach themselves to any hairy object with which they come in contact and there is a vastly greater number of chances that they will fix upon the wrong than upon the right insect. They have been found upon hairy beetles, flies and bees of the wrong kind. Those, however, which are fortunate enough to chance upon Anthophora are carried to the nest. Now the male Anthophora appears about a month earlier than the female, therefore, most Sitaris that arrive at their proper destination are attached to the males. They transfer themselves, however, to the female. When the female Anthophora lays her eggs in the cells of the nest the triungulin larvae slips off her body onto the egg she has just deposited upon the honey. Here the larvae remains, balanced carefully upon the egg, for if it left for the honey, it would be drowned. The bee then seals up the cell, and the larvae proceeds to eat the egg, living upon its contents for about eight days. It remains in the shell of the egg during this time for it would be suffocated if it came into contact with the honey. At this period the larva moults and appears in the form specially adapted to floating upon the honey, which is to be its food for the next six weeks. The legs of the triungulin stage have disappeared together with the other appendages and the larva now seems but little more than a vesicle. It is shaped, however, in such a way that one surface must float uppermost in the honey, and round this surface are the openings of the spiracles, so that the animal is adapted to breathe while it floats passively upon the surface of the honey, which is its food. When it has finished the honey it is metamorphosed into a pseudopupa.

Individual Sitaris may vary in their subsequent life history before reaching the stage of the perfect insect, but we will leave these stages and deal with those described. The main adaptations here are: the numerous eggs laid by the female, which meet the high mortality among the larvae, the three claws upon the leg, which enable the larva to cling to the bee; its emergence from hibernation at the same time that Anthophora appears; the instinct to leave the male bee and go to the female and to leave the female and float upon the egg; the equally remarkable instinct through which it rigidly keeps within the egg; the metamorphosis into a shape unknown among other beetles, which is perfectly adapted to a passive existence, floating upon the honey in the cell of the bee."

This shows the wonderful skill and intelligence of the cells in building the smaller individuals like insects. You will notice how the cells are able to change the form of the insect as it becomes necessary to effect the results desired. Sometimes he is this and again he is that, as circumstances require. In the past ages their knowledge and experience in life have taught them how to build all these different structures and perform all these different and difficult acts. Their memory of the past must necessarily direct them in every act. Those actions and experiences are their knowledge, that is to say, the cells that occupy the insect and build these different forms, cooperate in a social way and build these structures which are adapted and required in order to be able to rob the bee of its honey and convert it to their own use.

The various constructions produced by the cells show that they produce what they deem necessary and what may be required for their perpetuation and the continuance of life. They follow no fixed rule, but have an ideal and purpose, which one also finds in the productions of man. Can intelligent man equal the performance of this cell colony called Sitaris? I think not.

Mr. Binet has ventured so far as to show by their actions that they must be possessed of instinct, just as animals are and makes the following statements:

"We may reply upon this point, that there is not a single ciliate Infusory that cannot be frightened and that does not manifest its fear by a rapid flight through the liquid of the preparation.

"If a drop of acetic acid be introduced beneath the glass slide in a preparation containing quantities of Infusoria, the latter will at once be seen to flee from all directions like a flock of frightened sheep.

"Memory, according to M. Romanes, first begins with

the Echinoderms. Now Maebius, upon the occasion of a treatise upon the Folliculina ampulla, a ciliated Infusoria presenting complicated and interesting movements, properly remarks that every time an animal repeats the same action under influence of the same excitations, that fact proves that the animal is possessed of a memory. In fact memory is one of the most elementary of psychological facts.

"Lastly the primary instincts, according to M. Romanes, begin first with the larvae of insects and with annelids. We give in contradiction of this statement the recent observations of Verworn, which reveal the existence of curious instincts among the rhizopods. The Difflugia urceolata, which inhabits a shell formed of particles of sand emits long pseudopodia, which search at the bottom of the water for the materials necessary to construct a new case for the filial organism to which it gives birth by division. The pseudopod, after having touched a particle of sand, contracts and the grain of sand adhering to the pseudopod is seen to pass into the body of the animal. Verworn instead of grains of sand placed small fragments of colored glass about the animal; some time afterward he noticed a heap of these fragments on the bottom of the shell. He then saw a bunch of protoplasm issue from the cell, representing the new difflugia produced by division. Thereupon the materials collected by the mother-organism—the fragments of colored glass—came forth from the shell and enveloped the body of the new individual in a sheath similar to that encasing the mother. These fragments of glass loosely interjoined at first were now cemented together by a substance secreted by the body of the animal.

"Two facts are to be remarked in this observation: First—The act whereby the difflugia collects the materials for providing the young individual with a case is an act of preadaptation to an end, not present but remote This act, therefore, has all the marks of an instinct. Further the instinct of the difflugia exhibits great precision, for the difflugia not only knows how to distinguish at the bottom of the water the material available for its purpose but it takes only the quantity of material necessary to enable the young individual to acquire a well built case; there is never an excess.

"It is interesting to note that the difflugia does not act differently from animals possessing more highly complicated organization and endowed with differentiated nervous system, as for instance, the larvae of Phryganids, which form their sheaths from shells, grains of sand or minute slivers. We shall not regard it as strange, perhaps, to find so complete a psychology in the history of lower organisms when we recall to mind that agreeably to the ideas of evolution now accepted a higher animal is nothing more than a colony of protozoans. Every one of the cells composing such an animal has retained its primitive properties, giving them a higher degree of perfection by division of labor and by selection. The epithelial cells that secrete the nails and the hair are organisms perfected with reference to the secretion of protective parts. Similarly the cells of the brain are organisms that have been perfected with reference to psychical attributes."

The cells described here by Mr. Binet are those still living a single life in the water. This cell has the habit of making a case around himself from the material at hand. If no fine sand is at hand, he is able to use powdered glass. A large number of other cells use the lime and other material found in solution in water to make their shells or covering. The actions of these single cells are not instinctive as Mr. Binet suggests, they are conscious intelligent acts. The cell knows what he wants and what he is doing. As far as his experience in life has been, he has only gone so far as to make for himself a covering of some hard material. We must remember that he not only picks up such grains of sand or glass as he thinks necessary, but he knows how to change this material into forms and substances with which he makes a shell and covering for the young cell.

You can see that he must have a knowledge of chemistry and also of the right proportions of mixing the material with which he makes his armour. He must understand how to dissolve the crude matter and with it make a mixture that will harden into a shell after having first been formed into the proper shape. He must perform all this work, having in mind and in view placing therein of the young cell as soon as he gets the new structure completed. Every act in this performance requires the same purpose and foresight as similar actions in man.

Mr. Binet also gives the following interesting description of the action of the male cell in animals:

"Let us now follow the spermatozoid in its journey to the ovule. It is known that the road it has to travel in certain instances is extremely long. Thus in the hen the oviduct measures 60 centimeters, and in large mammifers the passages have a length of from 25 to 30 centimeters. We might ask ourselves how such frail and minute creatures come by a power of locomotion great enough to enable them to traverse so long a path. But observation discloses the fact that they are able to overcome obstacles quite out of proportion to their size. Henle has seen spermatozoids carry along with them masses of crystals ten times larger than themselves without appreciably lessening their speed. F. A. Pouchet has seen them carry bunches of from 8 to 10 blood globules. M. Balbiani has attested the same fact. These globules, which have fastened themselves about the head of the spermatozoid have each a volume double that of the head. Now according to Welcker, the weight of the globule of human blood is 0.0008 of a milligram; allowing that the spermatozoid has the same weight, we may then say that it is able to carry burdens four or five times heavier than itself."

This germ cell from the male looking for the female cell certainly shows activity and a will power. After considering the actions of the male cell, Mr. Binet also makes the following remarks: "Another remarkable circumstance is that the copulation of the two sexual elements is not without analogy to the copulation of the two animals from which they originated. The spermatozoid and the ovule to some extent repeat on a small scale what the two individuals perform in their larger sphere. Thus it is the spermatozoid that in its capacity of male element goes in quest of the female. It possesses, in view of the journeys it has to make, organs of locomotion that are lacking in the female and are useless to it.

"In fine, the spermatic element, in directing itself toward the ovule to be fecundated, is animated by the same sexual instinct that directs the parent organism towards its female."

This is only what we should necessarily expect when we consider the matter. The intelligence that guides the actions of man or animal in reference to their mating instinct is and must be the same as that of the germ cell. The actions of animals called the mating instinct are originated and instigated by these very germ cells; that fact is shown conclusively by the removal of the organ containing the cells.

It is now admitted, I believe, that habits are based on the power of memory in the cell or organs performing the actions. If the cells of the body possess the power of memory so as to be able to remember how to perform the complicated movements involved in playing a piano, guiding an auto, etc., it seems only reasonable that the germ cell should be able to remember how to build another body like the one from which it came, especially when he

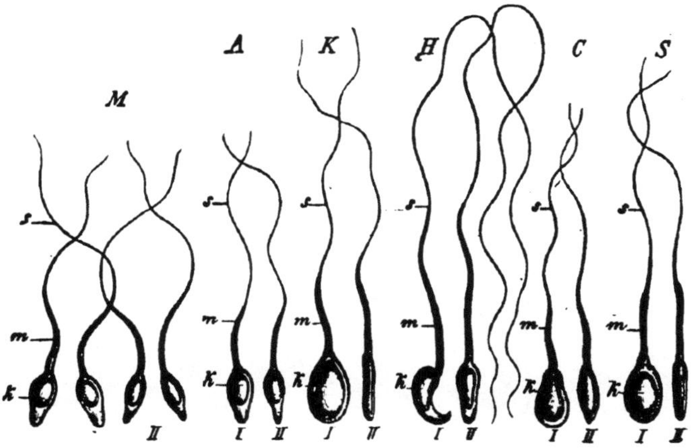

FIG. 43.—Seed-cells or sperm-cells from the semen of various Mammals. The broad side of the flattened, pear-shaped nucleus portion of the sperm-cell (the so-called "*head* of the sperm-animalcule") is represented in the drawings marked *I*; the narrow side in those marked *II* : *k*, kernel of the sperm-cell; *m*, central portion (protoplasm); *s*, active tail-like process (whip); *M*, four human sperm-cells; *A*, two sperm-cells of the ape; *K*, of the rabbit; *H*, of the common mouse; *C*, of the dog; *S*, of the pig.—SCHUTE.

has been set aside as a special committee to look after that work and nothing else. The cause of heredity, then, is the memory and intellect of the cell. Memory is the power possessed by the cell to take and keep a record of past events and experiences in such a way that they can be used and referred to in the future when necessary to guide the actions of the individual. Mr. Ribot states: "Some recent authors, among them Dr. Maudsley, attribute a memory to every nerve cell, to every organic element in the body. Dr. Maudsley states: 'The permanent effects of a particular virus, such as that of variola or syphilis in the constitution show that the organic element remembers for the remainder of its life certain modifications it has received. The manner in which cicatrix in a child's finger grows with the growth of the body, proves, as has been shown by Paget, that the organic element of the part does not forget the impression it has received. What has been said about the different nervous centers of the body demonstrates the existence of a memory in the nerve cells diffused through the heart and the intestines; in those of the spinal cord, in the cells of the motor ganglia, in the cells of the cortical substance of the cerebral hemisphere.'"

You notice that he calls the cell "Organic elements" and "Cortical substance," similar to Haeckel, who called them "Plasm" or "Living substance." There seems to be no one yet who seems to realize what the cell really is. They seem to think that because he is small he is merely a living matter or substance. Those who know his various activities seem never to stop to consider that he is a very highly organized and specialized being, and that he is possessed of a number of special organs, the purpose of which we do not know. While we see and know considerable about him, still our knowledge is very limited and very likely he has many powers and senses we yet know nothing about. We know practically nothing about the inner life of the primordial being of which the cell is composed. We may some day be able to produce a microscope powerful enough to be able to see the separate actions of these primordial beings that we now call cell granules, and if so, we shall likely discover the first cause.

We can see enough of the actions of the cell to demonstrate, I believe, those facts which I claim for him. Take for instance the actions of the single cell called Didinium, which is able to make darts or spears and throw them at its victim from a distance. Its actions are described by Mr. Binet in the following language: "These organs are the weapons used by the Didinium in attacking the live prey which constitutes its sole nourishment. Not only does it attack and devour animalcules almost as large as itself, but frequently it even seizes individuals of its own kind. In such cases it is always Infusoria and never the Rotatoria, although the latter often abound in waters which the Didinium inhabits. It appears, moreover, to have a marked predilection for certain species and so it happens that the huge and inoffensive paramecium aurelia is almost always its choice by preference among the animalcules that inhabit the same liquid.

The prehension of food by the didinium exhibits interesting aspects which have not as yet been observed in any other Infusory. M. Balbiani in his first observations had often been surprised at seeing animalcules that the didinium had passed by without touching, suddenly stop as if violently paralyzed; whereupon our carnivorous specimen straightway approached and seized them with seeming facility. More careful examination of the didinium's actions soon furnished the key to this enigma. If, while swiftly turning in the water, the didinium happens into the neighborhood of an animalculum; say paramecium, which it is going to capture, it begins by casting at it a quantity of bacillary corpuscles, which constitutes its pharyngeal armature. The paramecium immediately stops swimming and shows no other sign of vitality than feebly to beat the water with its vibratile cilia; on every side of it, the darts lie scattered that were used to strike it. Its enemy then approaches and quickly thrusts forth from its mouth an organ shaped like a tongue, relatively long and resembling a transparent cylindrical rod; the free, extended extremity of this rod it fastens upon some part of the paramecium's body. The latter is then gradually brought near, by the recession of this tongue shaped organ towards the buccal aperture of the didinium, which opens wide, assuming the shape of a vast funnel in which the prey is swallowed up."

We have here a single cell who has built around himself an armor hard as stone for his own protection in battle with other cells or enemies. He has made holes in regular rows around his body covering, through which he sticks hands, with which he paddles himself through the water. He also makes hundreds of darts or little spears, which he carries with him while hunting for other smaller cells and with which he is able to strike and kill from a distance his victims, which he devours as his food. He must be able to see and judge distances, or else how could he guess or know when his victim was within striking distance? He must be able to feel and taste or else how could he tell when he had hold of it or what he was eating, and so on, all through his different performances. This being is a cell. It multiplies by dividing in two, in the same manner as all cells, including the cells that build the human being and plants. In what manner do the actions of this cell differ from those of a human being as far as showing intelligence, considering simply the actions themselves, and not the size of the actor? In what manner do his actions in covering himself with an armor differ from the same actions by man? In what manner do his actions in making weapons and in hunting other animals for a living differ from those same actions performed by man?

Consider again the actions while mating of the cell living a single life, and those of the germ cells pursuing the female cells. Mr. Binet states that "the movements they execute admit of exact comparison with the actions attendant upon copulation among higher animals," and then illustrates their maneuvering in the following language:

"Upon the approach of the period for propagation, the paramecia come in from all points of the fluid and assemble like little whitish clouds in more or less numerous groups about the objects that float upon the surface of the water or adhere to the sides of the vessel, containing the tiny artificial sea in which the animalcula are held captive. Intense excitement, which the need of food does not suffice to explain, prevails in each of these groups; a higher instinct appears to dominate all these tiny organisms. They seek each other's company, chase each other about, feel here and there with their cilia, adhere for a moment or so in an attitude of sexual coition and then retire, soon to begin anew. When these minute assemblages are dispersed by shaking the liquid, they quickly form again at other points. These singular antics wherewith animalcula appear to incite each other mutually to copulation often continue for several days before the latter act is definitely affected."

All these actions of the cells that we have now been considering, whether single cells or social cells, like those who build animals and plants, are in general identical with those of the higher animals, including man, who we all agree are intelligent beings.

Before closing this chapter we shall examine a few cases which will illustrate more fully the fact that like does not always produce like as is generally understood, but that the cell will put up such structures as will be required and necessary under the particular circumstances. Take for example the life and development of a butterfly. The egg is laid in the fall and in the spring this egg, a single cell, multiplies and with the material furnished by the parents and at hand, such as the leaves of plants and trees, builds itself into a structure like a worm and moves about, filling this structure during the summer with more building material to be used later on; this worm or caterpillar is now a moving structure like an animal or automobile. The next thing that happens is that towards fall this moving structure changes into a chrysalis which is a stationary structure.

What is the purpose? It is for the purpose of changing later on into a flying machine. The worm or caterpillar takes part of himself and with this material builds a house or covering around himself; then he tears himself to pieces and rebuilds himself into an aeroplane and occupies the house first built as a work shop, while rebuilding himself into an aeroplane or flying machine. However, we must remember that it is not the worm or caterpillar that does these things, it is the cells of which he is composed and which put him together in the first place. They take the first structure, the worm, apart, tear it down as man would a house and with the same material build a new structure, a flying machine—the butterfly. Now you will notice that in this case as in a great number of other cases, like does not produce like. The worm does not produce a worm nor does the butterfly produce another butterfly. In this case, the worm produces a butterfly and the butterfly produces an egg. Their heirs are in no way similar to themselves. The fact is that no animal or plant produces anything, it is the cell that produces. In the same manner the submarine or house does not produce anything; it is the builders of these structures, man, that produce them. In the case of the caterpillar changing into a butterfly, the caterpillar is entirely destroyed as an individual, but the builders and material are still there and they take the same material and with it build a new and different structure calculated to move through the air. Mr. Walker states that there are various actions caused by the nourishment which they obtain. The following are his ideas about the cause: "Take the life history of a moth or butterfly, as an example. A caterpillar hatches out of the egg complete in all its parts and capable of looking after itself. Directly it hatches out it simply grows in response to the stimulus of nourishment. It does not require any knowledge with regard to the kind of food which is necessary for its well being; it is born with this knowledge, When the next stage in its life-history arrives, it spins itself a cocoon and that without ever having seen a cocoon. This knowledge and skill is an inborn character and the caterpillar is prompted to spin a cocoon at the proper time, purely by instinct. When in the cocoon, the caterpillar is metamorphosed into a chrysalis and in the chrysalis all the parts of the perfect butterfly or moth are developed simply by growth. When the butterfly or moth hatches out in due course, all its muscles and organs are developed and it is not dependent upon the stimulus of use, that is of exercise in the case of the muscles, for development." He states that it simply grows in response to the stimulus of nourishment and that the butterfly or moth is developed simply by growth. Mr. Walker has written what is considered to be a good up-to-date work on the cause of heredity, but I do not agree with him in the matter. He does not consider the cell anything but a chemical force. There is nothing in life that "simply grows." Everything is put together for a purpose and it is put together by someone. It does not come together by chance. Mr. Walker calls our attention to a peculiar plant and worm living together in partnership as one individual. The plant lives inside of the worm and inside of this worm the plant makes food and building material for both. He states: "A still more remarkable case is afforded by a low form of worm (convoleta roscoffensis) which lives symbiotically with green algae. The algae appears to form a special assimilation tissue within the worm enabling it to live like a green plant. The worms are elongated and colored green and at Roscoff live in the sandy tide pools fully exposed to the sun's rays, looking like a mass of floating weed upon the surface of the water. Now the stomach and indeed the whole of the alimentary apparatus has disappeared in the worm and it lives exclusively upon carbohydrates formed within its body by the algae. The algae has undergone most profound changes. It has lost its membrane, thus allowing its secretions and excretions to pass freely among the cells of the worm and it cannot live independently. If the worm dies the algae contained in it die. All the allied forms of this worm seek the dark and live concealed under stones and vegetation and they are carnivorous. Convoluta, however, being dependent upon the well being of the algae for its nourishment, seeks the sunlight and the surface of the water. The sunlight is necessary to the growth and metabolism in the alga, which in its turn supplies food to the worm, which has ceased to be carnivorous. The two forms are thus absolutely dependent upon each other and incapable of living apart."

I am quite certain that Mr. Walker and others who have examined this creature are mistaken in reference to the facts. I think it will be found upon further examination, that this cell colony is made by the same family of cells, and that it is not a co-partnership of a plant and an animal. This individual moves up into the sunlight where it can manufacture starch, etc., from the raw material found in the water by the aid of the sun's heat or rays. It is necessary to get into the sunlight in order to be able to make starch for food and the outside transparent mobile skin like a worm's is simply a structure made by the cells for the specific purpose for which it is used. There is no difference in the general appearance of a plant or animal cell. The cell can make either plant or animal and will make animal parts such as will be necessary for its use in its struggle for existence. They call it a plant cell when the cell knows how to make food and other building material from the raw material with the aid of sunlight. The other cells are supposed to live a parasitic life and to depend on the plant cells for their food and building material. However, the distinction is not reliable nor of any consequence, as we find the plant cells also are at times carnivorous and live on insects and other similar food, as for instance in the case of the pitcher plant.

Before closing this chapter I wish to again call the reader's attention to the fact that the old idea of the law of heredity, that like produces like, is of no particular significance, but that the cell will produce such structures or habitations as they think will be the most useful and necessary in every particular case, and even change those already produced, if they think best.

Take the case for instance of the bee. Here we have males, females and neutrals or workers. The female lays the eggs that produce all these three kinds of individuals. Mr. Haeckel states: "In the bees we have the remarkable feature that it is only decided at the moment of laying the egg whether it is to be fertilized or not; in the one event a female and in the other a male bee is formed from it."

The egg is a cell and it knows how to build any one of these three individual structures. The workers indicate what is wanted and when the cell is placed in the comb it is told by the cells in the Queen what is wanted. The cells of the female bee know at all times by being in close touch with the crowd and the colony just what is most needed in the community and every cell as it is placed in the comb will know what it has to do, and in this manner the swarm is constantly kept in proper proportions. There is nothing more strange in this case than in the case of the butterfly, or in the building of an animal. After the stomach of the animal has been started the cells first begin to gather in crowds here to start the head and there to start the limbs, etc., and unless they understood each other perfectly, they could not possibly go ahead with their work in perfect harmony and understanding and complete a plant or animal. The next and last illustration I shall give in this chapter on heredity is that of the social ants, in order to fully illustrate that there is a guiding mind and intellect which directs the course of development of plants and animals. In many of these classes of social ants there are as many as five different kinds of individuals, such as males, females, workers, soldiers and slaves; and in one case, some of the insects turn into storage tanks in which the honey is stored and as such are suspended under the roof of their dwellings. Mr. Walker has the following to say in reference to some of these social ants:

"In some ants such as the driver ants of Africa (Typhlopone) the physical differences between the queens and neuters is so great that even trained entomologists have been mislead and have classed them as entirely different insects. Thus the queen, the male and the worker of Typhlopone are now known, but until comparatively recently the queen was put in one genus (Dicthadia) the male in another (Dorylus) and the worker yet in another (Typhlopone) and this mistake was made by skilled entomologists, who were misled by the enormous difference in the structure of the three kinds of individuals. The male has well developed eyes while the worker has none.

"The polymorphism among social ants and bees has been the subject of much argument and has produced a great deal of speculation. We are not here concerned with such questions as how polymorphism has arisen. It is sufficient to emphasize the fact that the potentiality of producing the various morphological characters peculiar to the different forms of workers, to the males and to the queens, must be present in the eggs of the queen, though neither she herself nor any direct ancestor has possessed those that appear regularly in the workers. Among the honey ants (myrmecocystus) some workers are used by the others as reservoirs for honey. These individuals remain in the nest clinging to the roughened ceilings of certain galleries. In the other galleries the ceilings are smooth. They remain in this position for the rest of their lives. Beforehand they feed on honey for some time and when they have eaten as much as they can, take up their positions in the nest. Here the other ants feed them with honey until they are distended to many times their normal size. Honey is not procurable at certain seasons of the year and the stored honey is regurgitated by these individuals as it is required to feed the larvae and workers. It is obvious that nothing of this kind can ever happen to the queen.

"Nor are our instincts the only characters transmitted through individuals that do not possess them. The same thing happens with regard to very striking physical characters. The soldier ant in a case of true ants is an example. The soldier ants are neuters and in them the head is greatly enlarged as also are the mandibles. Various parts of the body coverings are modified and serve the purpose of defense. In fact, the soldiers of the colony are so different from their parents, the males and queens, that the untrained observer would class them as belonging to a different family."

Now it is in this case perfectly clear that like does not always produce like and that the males and females produce animals that are not only unlike themselves in size and general appearance but unlike in their ways and habits. The workers and soldiers produce no offspring, so their characters cannot be transmitted to posterity through heredity in the sense in which it is generally understood and which with the evolutionist is the cause of species. There cannot be any more of a mystery in this case than in the case of the development of a plant or animal from one single cell. The individual that is hung up under the ceiling and used by the others as a storage tank for honey surely cannot transmit any of his form and characters to his heirs or posterity, as he never has any nor is it intended that he should. From our point of view it would seem cruel punishment to be suspended under the ceiling and treated in this manner, but when we consider it rightly from the standpoint of the builder, the cell,—and that someone must be in charge of the food, we can see that the individual cells which together make up this storage tank for honey are alloted just as easy work as any of the others. Someone must be in charge and take care of the honey, for many kinds of single cells such as bacteria or fungi would eat this honey if permitted, and it is a wise thing to have it taken care of in this way. It is no punishment to the colony of cells known as the caterpillar, to reconstruct themselves into a butterfly, that is, from an ordinary moving structure or habitation to a flying structure, nor will it be any punishment to this colony of ant building cells to change their moveable habitation into a stationary structure where they can live in comfort and eat all the honey they please.

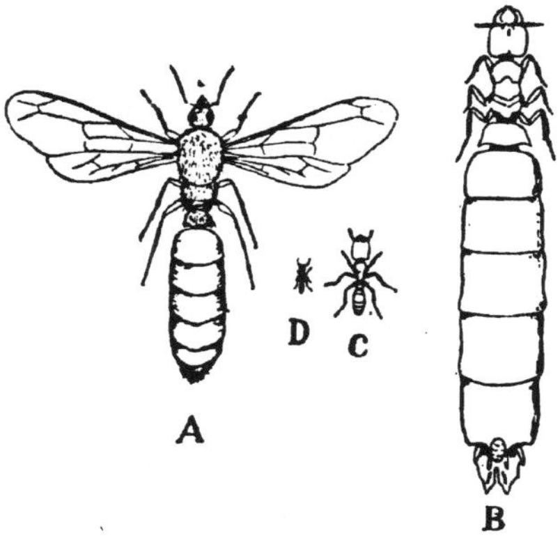

FIG. 44.—Ants. *A*, Male of *Typhlopone*, formerly classed in a separate genus (*Dorylus*). *B*, female, formerly classed in a separate genus (*Dicthadia*). *C*, Neuter soldier (worker). *D*, Neuter minor (worker). [Actual size.]—WALKER.

In this case of the social ants, it is clear that the cells of the males and females are in close touch with the whole social community of ants, and as each egg or germ cell is laid it is instructed as to general conditions and as to what is wanted or needed in the colony, and upon such direction the germ cells act and produce a worker, soldier, male or female, according to the order from the outside through the parents. That the cells receive and act upon communication from the outside is shown in many ways, as in the case of the flat fishes where a picture of the outside color and formation is transmitted from the eye to the skin cells. For instance, the skin cells of the fish will change both the color and shape of the skin to simulate the shape and color of the rocky bottom of the sea on which he is resting. Just as in the case of the social ants, the one and the same germ cell of the animal can build different kinds and different shaped organs, as head, tail, stomach, hair and limbs. All these five structures are entirely different and for a different purpose. The cells must be told in some way what to build and be able to communicate and understand each other very clearly or else they could not possibly work in harmony, and have no one interfering with the other, but each producing just what is required and nothing more.

The seventeen-year locust lives in the earth seventeen years as a worm; then he tears down the seventeen year old structure and with the same material he builds a flying machine, a new and entirely different structure, which is known as the cicada. This flying machine or cicada is produced for the one special purpose of finding a mate and thereby perpetuating his kind. This flying machine lives but a month or two. At first he lived the life of a worm and was a worm but, as a worm, he had no offspring and no one could inherit his character. But when we understand life correctly, that all living things that we see are structures produced by an intelligent being for his special purpose, we can clearly see why like does not always produce like and that living structures are produced only when wanted or considered necessary by the builders.

CHAPTER 8.

CAUSE OF INSTINCTIVE, EMOTIONAL AND REFLEX ACTION.

Life is both form and action, but the more important part is action. I have not been able to find anyone who has understood the cause of instinctive, emotional and reflex action. The reason for this it seems to me, is because no one has ever been able to comprehend and look upon the cell as an intelligent being. The old phraseology and idea that only man can reason and that all animals act from instinct is to blame for it. One side claims that the intelligence back of instinctive acts of animals is God or nature and the other side does not claim to know. Take for instance, the instinctive actions of the young pig. The instant he is born he is able to go to work and obtain nourishment from the mother. Without having had any previous experience or training he knows how to do those complicated acts as if he had been practising for years. These actions are called instinctive. They are intelligent acts but still according to the old ideas we cannot consider any actions intelligent unless performed by man based upon previous experience and training. There are a great number of actions performed by animals and plants and man that are performed without any previous experience and are called instinctive. The instinctive actions are similar to those performed without our consciousness such as we call habitual, reflex and automatic. Now it is clear that these actions are all directed by someone. Who is this someone? In the submarine or battleship it is man, the builder of these structures, who directs their movements; so it is with organic individuals we call plants or animals, it is the cell, their builder, who directs their movements.

It would be just as consistent to say when looking at a battleship or submarine from a distance, not being able to see the man in charge, that the battleship or submarine had an instinct to shoot and hit a target at a distance, without having had any previous training. The swallow can fly the first time it makes the attempt. It makes no attempt until it is fully developed and ready. So with the aeroplane, which can fly the first time it makes the attempt because as is evident and natural, those who know how to build the machine also know how to operate it. Following are extracts from Hallock's Psychology, defining the different kinds of actions:

First—"Unconscious, reflex action—such as the unconscious movement of the sleeper's hand when touched.

Second—"Conscious, reflex action—such as winking the eye, due to sensitiveness to light. This is illustrated by a sensation which reaches the brain and hurriedly passes out in motor action.

Third—"Impulsive action—Here a hazy idea of a purpose toward which the action tends makes its appearance.

Fourth—"Instinctive action—This cannot always be separated from impulsive action.

Fifth—"Deliberate action.—Here a deliberate Will chooses between alternative courses of action. Shall I spend this money for books or several other things?

"It is sometimes hard to believe that reflex actions are not consciously willed. We have seen that a decapitated frog raises its hind foot to brush away a needle or a drop of acid on its side; if the leg on that side were amputated, the other leg would display purposive endeavor to remove the object.

"It is difficult to appreciate properly the complexity of reflex action. Many movements which seem to exhibit all the purposive guidance of the deliberate will are nothing but reflex actions, which were at first consciously willed and often sink to the level of reflexes, such as walking, writing, balancing and many other muscular movements.

"Laws of Central Nervous Action—Whenever a sensory stimulus is transmitted to a central nerve cell, the force is never lost but intends to flow out again in motor action. By central nerve cells we here mean ganglia in either the brain or the spinal cord. Whenever a sensory stimulus pours into nerve cells, there will be a tendency for it to pass out in motor action, which causes muscular movement. This law holds equally good for conscious or unconscious stimuli.

"No reflex action shows the presence of will in its higher deliberative form, although purposive reflexes may indicate a time when they were the slowly formed products of all the individuals, intelligence and will power.

"Instinctive action—Whenever a conscious sensation, due either to external or internal stimuli results in purposive action toward a given end, (which is not forseen) that action is instinctive.

"A young stork left alone in a northern latitude would emigrate southward on approach of winter; had the bird never been south before, it could have no idea of the purpose of its flight although its actions were directed toward an intelligent end. Certain sensations of organic origin prompt the young bird to build its first nest. This bird has never been taught nor has it had any experience of nest building, yet the first nest is constructed on the same plan and fashioned as well as any of its successors, all the actions, gathering of straw, grass and twigs, are a series of complex movements blindly directed toward an intelligent end.

"Instinctive compared with reflex actions—Instinct has much in common with reflex actions. Whenever certain stimuli are present, definite, unvarying actions tend to result in both cases. We saw that a sensory stimulus on the side of a decapitated frog was followed by a definite action suited to remove the cause. When the caterpillar feels certain stimuli, it mechanically begins to weave a shroud in a blind reflex way and the action is automatic as long as the stimulus is operative.

"A reflex movement is simpler and does not involve the whole body in action. A limb may be moved; an eye winked; one muscle contracted. When a bird builds a nest, the instinctive tendency results in movement of wings spread in flight, ocular search for materials, alighting and seizing them with either bill or claw, carrying them to proper place and fixing them in position. Here the instinctive movements constantly change and the one is not a mere repetition of the other.

"An ant will hoard grain for the winter and the caterpillar provide for a butterfly existence.

"Bees construct larger cells for young Queen bees and feed the Royal Larvae with more and richer food, although there is originally no difference between them and the Larvae of Workers. Similar reflex tendencies would result in making all cells the same to start with and feeding all the young the same way. Hence, some call all instincts examples of 'lapsed intelligence,' that is the actions were at first the result of a highly voluntary process, but from a continued repetition they became unconsciously habitual.

"In the South Sea Islands, a species of bird accustomed to build nests on the ground placed them in trees after cats were introduced.

"Instinct urges the bee to labor hard to gather honey against the time when the flowers will be gone.

"After instinctive action has been once performed a certain amount of foresight of the end must remain in memory.

"The hen that has once kept eggs warm until they are hatched must have a certain idea of the results of the action when she again sets. By the law of contiguity the association would run straight ahead to the chickens, but this cannot be said of the first performance of any instinctive action, nor can it ever be maintained in such cases as when the animal dies or changes its form after an instinctive act.

"The silk worm never has an idea of the end in weaving its shroud."

The reader will notice that all these actions described by different animals will seem very mysterious and inconsistent if you do not stop to consider the situation correctly. You will admit that all the actions of these insects and animals seem intelligent but yet you cannot see how they can be, as the animals have had no experience, training or instructions in the matter. These actions have been called "Lapsed intelligence" meaning that they must have been intelligent at some past time, but when you stop to think of it how could they ever have been more intelligent at any other time? Did the hen at one time in the past possess the intelligence of man, God or the Creator?

When we talk about the peculiar actions of plants and

animals without taking into consideration the individuals that build and maintain them, we get all mixed up, and our statements become a jumble of meaningless words and mysteries. These actions—reflex, automatic, impulsive, emotional or instinctive, are all intelligent acts.

What we call a reflex act, like winking the eye, requires just as much intelligence as any other act. However, the intelligence is located in a bunch of cells in the brain that have that particular business in charge, and the other brain cells such as are in charge of receiving sensations from the outside world through sense organs, like the eye, ear, etc., do not have to be bothered with such details. Those cells that are in charge of receiving information from the outside world are what we call our consciousness. You can clearly see that it requires conscious intelligence, to be continually on the watch out for dangerous materials and to keep them from getting into the eye. The eye requires constant and careful attention and the wonderful manner in which it is taken care of by the cells in charge amidst the many dangers in life, shows loyalty and intelligence. The cells in charge of receiving information from the outside world must attend to this business of receiving, and they cannot be bothered with details like balancing, walking and other muscular details. These actions are for that reason delegated to other cells which we call nerve centers. These actions, when performed by us without any knowledge or consciousness, we call reflex, habitual, or automatic. These are simply names for something that so far has not been understood. Writers on psychology have not been able to comprehend that just as much conscious intelligence is required in one place as in another; that the cells in charge of one line of work are just as smart as the cells in charge of work in another place. The cells in the spine that look after the balancing and the action of the heart, must be just as intelligent as the cells which are located in the head and direct the actions of your arm or hand when you are learning how to perform some particular action like writing or balancing.

Instinctive actions in animals are of course the same as those in man. We are perfectly conscious of the act and still we feel as though something gets possession of us, as for instance when the mating instinct takes possession of us. We are then in love as we call it. The act of perpetuating the race is the most important thing of man's life and it also requires the combined acts of the whole individual. For these reasons the cells see fit to give both mental and bodily control to the nerve center or cells who have that matter in charge. No individual will act except in accordance with his thoughts, so it is clear that in all instinctive actions in man and animals, insects or plants, the actions are the will of the party who performs them.

The following by Mr. Hallock shows some instances of animal intelligence: "A man allowed a sow pig about a year old to run in his orchard. He watched her go to a young apple tree, shake the tree and eat the apples which fell. Having finished these, she again shook the tree, pricked up her ears and listened for more to fall, but as none fell she went away.

"Another sow, with a litter of pigs, was accustomed to spending the day in a forest, returning home at night to be fed. When her pigs were of sufficient age, three were taken to be roasted, at different times, being caught when she returned home with them. After the third had been taken, she came without the pigs. The next evening a watch was set to find out what had become of them, but she would not allow them to follow her farther than the edge of the forest and drove them back repeatedly; she then went to the house to get her own supper and returned to them. She had evidently discovered their disappearance on going to the house, and took this course to save them.

"A man once desired to test his dog, which was lying down quietly sleeping. In the midst of a general conversation, he inserted the words, 'The cow is in the potatoes.' The dog jumped up at once, dashed out to the garden and appeared surprised at not finding the cow there.

"Another dog had for some time chased a rabbit, which ran in a circular course to a burrow and escaped the dog. Finally the dog, on starting another rabbit, ran indirectly across the circle to the mouth of the burrow and awaited the rabbit there.

"It is well known that some ants keep a certain species of insects called aphides. These stand in precisely the same relation to ants that cows do to human beings. The aphides are regularly milked by the ants and a sweet nutritious liquid somewhat resembling honey is thus secured. The ants sometimes build stables for their cows, allowing them to pasture on certain plants in such a way that the cows cannot escape. Naturalists have repeatedly seen ants milk their cows.

"Three species of ants keep other ants as slaves. The slaves tend the aphides, milk them, and often climb trees and plants in order to find male aphides to increase the dairy. The slaves even put the food into the mouths of their masters. Bugs are also enslaved by ants and made to carry heavy burdens.

"Ants have been seen to make a bridge across a small vessel of water, using pieces of wood and straw. The ants might have brought earth for this bridge, but they illustrated the choice of means to ends, and used instead pieces of wood and straw.

"Ants have cemeteries and funeral processions. Two ants bear the corpse at the head of the procession, which is followed two by two. When the first two are tired the next two take up the body and on arriving at the burial ground, they dig a grave and inter the body. A lady, wishing to see a funeral procession, killed a number of soldier ants. She watched the procession as above described. On reaching the cemetery, six or seven of the ants refused to help dig the graves. These ants were caught, brought back and killed at once, like deserters from an army. A trench was then dug, in which they were all buried together. It was observed that ants would not bury their slaves in the cemetery used for the masters. Darwin called the brain of an ant one of the most marvelous atoms of matter in the world."

The dog that ran to the potato patch knew from association with the expression "cow" and "potato patch" what was needed or wanted, he remembered it. From the dog's recollection of his experience with the rabbit, he decided that it was useless to try to catch him as he would beat him to the hole, so he decided to try this new experiment and take the rabbit by surprise.

In hunting jack-rabbits, every one knows the trick of the jack, to always turn square to one side as soon as he goes over a hill and out of sight. This trick will always fool both man and dog.

The pig could observe and remember that when he shook the tree an apple would fall. His actions were based on previous experience fixed in his memory. The pig reasoned from experience just as man would. The animals, like man, are able to reason and learn from experience. In their place in life they are just as intelligent as the German Emperor or the Judge of the Supreme Court of the United States.

In order to give the reader a more comprehensive view of the perfect, intelligent organization of the individual cells that build and direct the actions of the animal and man, I will give Mr. Hallock's description of the brain and nerves of the body. He says:

"Suppose that a child of intelligent parents was ushered into the world with perfect brain, with no optic nerve to transmit the glorious sensations of the eye, no auditory nerve to conduct the vibrations of the mother's voice, no tactile nerves to convey the touch of the hand. Could such a child live? No matter how perfect might be the child's brain and body, his faculties would remain shrouded in darkness. Perception could give memory nothing to retain and thought could not weave her fabrics without materials.

"It is the business of the nervous system to transmit the effects resulting from internal or external stimuli. This function of reporting stimuli may be compared with the machinery of an associated press agency, which gathers news from the world. The manager may be sitting in his office in New York or London and he cannot see what is taking place in the rest of the world, but there is a click of the telegraph instrument, and he learns that an ocean steamer has been wrecked on the Irish Coast. Another instrument vibrates with a message that a certain city cannot be heard from. The manager himself sends a dispatch for news and he now illustrates the second capacity of the nervous system, that of transmitting commands by its own peculiar automatic power; but he telegraphs in vain, for the wires leading to the city are broken.

"These telegraph wires are analogous to the nerves of the ear, eye and other senses. It is the business of these nerve cells to report what is taking place in their own special world.

"The brain cell in its dark chamber can receive dispatches from them alone. If a man loses his sight, the optic nerve brings in no further news and the case is similar to that of the distant city whose telegraphic communications have been broken by an earthquake or otherwise.

"The nerve cells traverse every region of the body, just as telegraph wires thread a continent. Without some such method of transmission, the sensations of sight, hearing, touch and other special senses could as we have seen never reach the brain, neither could commands, such as to move the muscles, be sent out from it.

"There are nerves which regulate the size of the blood vessels and the nourishment of the body, control the secretions and perform various other offices connected with transmitting stimuli, which are neither sensory or motor.

"A ganglion is an aggregation or group of nerve cells. Each ganglion is in some respect a little brain. The spinal ganglia receives a sensory impulse and sends out a motor dispatch without calling on the brain. The spinal cord is largely made up of nervous ganglia, sometimes called the little brain.

"If one were to prick the toe of a sleeper, the sensory nerve at that point would report the fact to one of the lower masses of nerve cells or ganglia, without waiting to hear from the brain; it would issue a command to the motor nerve cells and the foot would be immediately withdrawn. Unless the thrust was severe, the sleeper would not awaken, nor would he be conscious of pain or the movements of the foot.

"Thus the cells of the brain are not only saved the

trouble of attending to every little movement, but much time is gained.

"After the child has learned the difficult art of balancing himself on his feet, walking is largely attended to by other nerve cells than the brain.

"If acid is placed on the side of a decapitated frog the animal will bring its foot to the spot and try to brush the drop away. Man also has this power after death.

"The pectoral muscle of a beheaded French criminal was pinched and the right hand was raised to the spot, as if to remove the cause of the injury.

"The medulla oblongata at the upper end of the spinal cord has more or less control of the sympathetic nervous system, which regulates the heart, lungs, blood vessels and various abdominal organs. If the muscular action of the heart were under direct mental supervision, a person might become so interested in something or so excited by an accident or unusual event that he would forget, until it was too late, to move the muscles of the heart. The same would also be true in respiration. The medulla has power to attend to these, without calling on the brain and obtruding the unnecessary details on consciousness.

"The brain, like a large city, has much of its complex business systematized and localized. The senses report to certain parts of the brain, while other well defined parts send out a motor order to raise a hand or speak a word.

The motor zone, or that part of the brain concerned in sending out orders to move the body, lies on either side of the fissure of Rolando. So definitely has this area been mapped out that it is possible to find for the purpose of a surgical operation so small a center as that which moves a vocal cord, directs a thumb or winks an eye.

"Sensory brain tracts are those concerned in receiving

impressions from the senses. The center of sight is in the occipital part of the brain, that for hearing in the rear, two-thirds of the first and second temporal convolutions.

"The Senses:—If currents from the various sensory nerves did not flow into the brain, we should get no knowledge of the outside world. The brain gets despatches from the optic, olfactory, tactile, gustatory and other nerves. These despatches are the data from which we get our knowledge of the world."

Now just consider for a moment what this means! What conscious intelligence and foresight is required to be able to effect a social organization like this! Man has been able to effect a similar civilization only in the last few centuries. The cell has understood the power of organization for a million years before it was conceived by the mind of man; still man as an individual pretends to be more intelligent than the cell, his maker,—like a submarine trying to be more intelligent than the man who put it together and directs its actions, or like a house knowing more than the architect that put it up.

Th. Ribot states that instinct and intelligence are one and the same thing except that "Instinct is unconscious intelligence," but Mr. Ribot does not understand what this unconscious intelligence can be. There are several books written about the unconscious intelligence and no one seems to know what nor where it is. It is perfectly plain that the acts we call instinctive and reflex are those performed and directed by those nerve cells of the body whose specific function it is to direct and perform those acts. A child can suck and make noises; he has had no previous experience or training in this matter, but he knows how. His mouth and throat were made for these specific functions. The makers are in charge, directing the different complicated movements of these structures and they know how to direct the work for which they were made. They know how to adjust the tension of vocal cords so as to make a noise when the air is applied.

A kitten that has never had his eyes open, or ever had any experience with a dog or a man, will show all the symptoms of terror and fright if placed near a dog. How does he know that dogs are dangerous, not having had any experience with them?

We say that instinct tells him. That is not true. Instinct is but a word expressing certain kinds of actions. Some one must tell the kitten and direct the kitten in its actions. Who is this some one? It is the builder of the kitten.

If the English ship should meet a German submarine in its first trip out at sea, it would act just like the kitten; although the ship had never met a German submarine before, it knew what to do because it had a crew in charge directing its actions who well knew that it was a dangerous enemy to the English ship.

When you consider who is in charge and directs the actions of the kitten, they are no more of a mystery than are the actions of the English ship, showing signs of fright upon the approach of the German submarine. Experiences of the past thousand years with the dog has taught the cells that he is a dangerous enemy.

We gave a name to this unknown cause and called it instinct, nature, providence or destiny but we know the cause of these instinctive, emotional and reflex actions, and all these actions go to prove that the cell is an intelligent being.

A text book on physiology describes the body in the following language: "The human body viewed from a mechanical aspect is the most perfect piece of work that it is possible to imagine.

"Most of us are willing to accept the facts that we can see, hear, taste, obey the impulses of our wills and think on what has past and what is to come without ever giving a thought as to how it happens that our bodies are so equipped as to enable us to do these things.

"The great foresight and preparedness of the German Empire has been the marvel of the age, but it sinks into nothingness when compared with the extraordinary equipment that the human body is possessed with."

The mind cells of man occupied with other matters have not time to comprehend all the details that are taken care of in the human body. The details are as vast and numerous as those of the largest city or empire, and are based upon the experience and intelligence of the ages while man has only begun to co-operate and organize in a social way. The reason we have never really understood causes in nature is our inability to overcome our old prejudiced idea, dished up to us in all our school books, that only man is intelligent. Think of all the complicated actions and movements of a chick! He is provided with an instrument with which to break his shell, and at once he is able to run about and pick up grains of sand, like his mother. Think how many extraordinarily difficult and complicated movements he is able to perform the first time without the slightest previous experience! He is not only endowed with an inborn dexterity in motion, but has also a powerful perceptive faculty. Without hesitation he picks up a kernel of grain. To do this he must be able to see and judge correctly and move his head and limbs with great precision. He could learn nothing in the shell.

We must remember, however, that but a month ago

the builders of this chick had taken part in the actions and were a part of the parents from which he came. Not only did this germ cell remember how to build the structure from which it came, but also how to maintain and operate it. The actions of the chick are called instinctive, but they are in fact no different from any other intelligent acts. The building, maintaining and directing of this structure we call a chick has been performed by this particular kind of cell for ages. His knowledge and experience is limited to this kind of structure.

The bee constructs a cell, the birds build a nest, the worm changes to a butterfly. These are all acts that the bird, bee or caterpillar have not had the slightest experience in or information of from the outside world.

In the case of the caterpillar and the butterfly, there is no chance to say that the caterpillar builds anything, because he is destroyed, gradually torn down, and rebuilt into a butterfly, a new and different structure. The cells or occupants of this caterpillar are all there, and so is the building material, and they change it into a flying machine.

The worm or caterpillar is in the hands of a power that can change him into a new and different individual. Think of the extent and magnitude of the operations performed in this task of tearing down the worm and rebuilding him into a butterfly. It must be admitted that it can not be done without a perfect system and plan, skillfully pursued and executed.

The cause of instinctive actions has seemed so mysterious to people in general, that preachers have used it in their sermons to prove the existence of a God, claiming that God is the instigator of all instinctive actions. I read the following Sunday Sermon from a Daily Paper: "It is not intelligence in the animal, nor of it. Back of instinct is law and back of law is the lawgiver. And who is that lawgiver but God? The intelligence of instinct is the intelligence of God. The laws of nature are the thoughts of God; and if instinct be only the outcome of law, it is the direct product of God's thoughts."

"Intelligence of all kinds comes from God," according to Dr. John P. D. John, former dean of DePauw University, who addressed an audience last evening in Grace Episcopal Church. He spoke on the subject, "A Glimpse of God in Instinct."

"The horse sometimes reasons; likewise the dog, cat and various other animals," he said. "An animal makes use of reason when it profits by its own experience. There are three kinds of animal actions in which means are adjusted to ends. These are reflex, instinctive and rational.

"Reflex actions are beyond the control of the will and are generally beyond the sphere of consciousness. The beatings of the heart and the chief processes of digestion, circulation and respiration are instances of reflex action. The intelligence in these actions does not belong to the animal itself.

"Instinctive actions are under the control of the will and the animal is conscious of them. All the animals of the same species under the same circumstances do the same things in the same way. They work according to a plan, but they do not perceive that it is a plan. Each animal does its work just like every other animal of its species, and just as its ancestors have done and as its posterity will do afterward. This class of actions embraces the greater part of the volitional work of the lower animals, such as the cell building of the bees, the nest building of birds and other forms of animal industry. In these cases the animals seem to proceed consciously to adopt means to ends, but they do it blindly. In this case of actions the animal does not benefit by its individual experience. It strikes out on no new path. Rational actions are those in which the animal profits by its own experience and strikes out on new paths.

"Numerous instances were given by the speaker as illustrations of the different kinds of actions. There is intelligence manifested in each kind of action,' he said. 'The intelligence of the reflex action is obviously not that of the animal itself, for it is not conscious of the action. The intelligence of the rational act is evidently that of the animal itself, for it profits by its own experience.' "

I agree with this minister in that these instinctive acts are intelligent, but not that they are God's actions. If God is back of the different instinctive acts in this world, I would consider him a bad actor. The mating instinct in man causes him to murder his fellow man sometimes upon the slightest provocation. His extra muscular development arose only from the mating instinct and we know from other animals of his kind that such muscular development is only for the purpose of contending with his fellow man for the possession of the female. His mating instinct drives him against his own reason to murder and suicide. This instinct will compel him to run away with his wife's sister and leave his own wife and family or to run away with another man's wife, while he has a wife of his own living. You remove the cause of this mating instinct and those instincts and troubles do not arise.

The instinct of the cat is to play with a mouse until it is dead, and sometimes to keep up the torture of the little nervous, sensitive creature for hours. This instinct to torture the little innocent mouse is unjust and unreasonable. The tiger will do the same with a child. The claws of the animal are made for that purpose. I could go on indefinitely and describe instincts of this character which would impute to God the instigation of actions most cruel and unjust.

The suffering in the world has been pointed out by philosophers and naturalists. Saint George Mivart says: "The world not only suffers, but has suffered for millions of years ere man was. For untold ages bloodthirsty rapine has raged and reigned, and cries of pain due to cruel wounds and to limbs crushed in bloodstained jaws, have continually resounded in the only one of God's worlds we are able to know and understand. The very existence of many creatures is bound up with the sufferings of others, and parasites, external and internal, torture their helpless and involuntary hosts, by implements carefully contrived for securing their hold and aiding their progress." All these different cruelties they call instinctive actions in animals and plants, yet I do not believe they are God's doings.

Bergson has the following to say about instinct in insects: "When the horse fly lays its eggs on the legs or shoulders of the horse, it acts as if it knew that its larvae have to develop in the horse's stomach, and that the horse in licking itself will convey the larvae into its digestive tract. When the paralizing wasp stings its victim, in just those points where the nervous center lies, so as to render it motionless without killing it, it acts like a learned entomologist and a skillful surgeon. But what shall we say of the little beetle sitaris, whose story is so often quoted? This insect often lays its eggs at the entrance of the underground passages dug by a kind of bee, the anthophora. Its larva, after long waiting, springs upon the male anthophora as it goes out of the passage, clings to it and remains attached until the "nuptial flight", when it seizes the opportunity to pass from the male to the female and quietly waits until it lays its eggs. It then leaps on the egg which serves as a support for it in the honey, devours the egg in a few days and resting on the shell undergoes its first metamorphosis, organized now to float on the honey it consumes this provision of nourishment and becomes a nymph, then a perfect insect.

"Everything happens as if the larva of the sitaris, from the moment it was hatched, knew that the male anthophora would first emerge from the passage. That the "nuptial flight" would give it the means of conveying itself to the female, who would take it to a store of honey sufficient to feed it after its transformation, that until this transformation it could gradually eat the egg of the anthophora, in such a way that it could at the same time feed itself. And all this happens as if the sitaris itself knew that its larva would know all these things.

"The yellow Sphex, which has adopted the cricket for its victim, knows that the cricket has three nerve centers, which serve its three pairs of legs, or at least it acts as if it knew this. It stings the insect first under the neck, then behind the pro-thorax and then where the thorax joins the abdomen. Is it not plain that life goes on to work here exactly like consciousness, exactly like memory?"

Mr. Bergson is compelled to admit that the actions of these insects show the same intellect as the actions of conscious man. However, we find that nearly all writers now agree that there is no difference between instinctive and intelligent acts. I maintain first that intelligence and instinct are one and the same thing; second, that the cells as well as animals and plants show the same instincts.

If you transfer any black bee to Australia or California,

the habits will completely change. She will find out that the perpetual summer will always provide honey in abundance and she will be content to live from day to day and just gather enough for each day's consumption. The same experience has been observed in man; neither man nor bee will hustle and store up provisions for the future unless compelled to do so. The cells in the brain of the bee and man direct similar actions.

Coming back to the different kinds of instinctive actions of animals and insects, I think that protective coloration is the most interesting and significant. I quote the following from Mr. Loeb's Mechanistic Theory of Life:

"That vision is based on the formation of an image in the brain, is supported by a group of facts. Sumner has shown that certain fishes are able to reproduce on their skin rather complicated patterns, to-wit:—a chess board, which forms the bottom of the aquarium. Panchet many years ago showed that the adaptation of fishes to the ground ceases as soon as their eyes are removed or as soon as the formation of retinal images is prevented through the turbidity of the refractive media of the eye. This fact proves that the so-called adaptation of fishes to their surroundings is only the transmission of the retinal image to the skin.

"It has been shown that the destruction of the optic fibers and the optic ganglia in the brain acts like the extirpation of the eyes and finally it has been proven that the cutting of the sympathetic fibers which go to the pigment cells of the skin also prevents the formation of the picture of the ground on the skin.

"Hence, we know the path by which the retinal image is transferred to the skin of fishes. The mechanism and conditions for the change in coloration was made clear by the investigations of Keeble & Gamble on the color change in crustaceans. According to these authors the pigment cells can, as a rule, be considered as consisting of a central body from which a system of more or less complicated ramifications or processes spreads out in all directions. As a rule, the center of the cell contains one or more different pigments which, under the influence of nerves, can spread out separately or together into the ramifications. These phenomena of spreading and retraction of the pigments into or from the ramifications of the pigment cells, form on the whole the basis for the color changes. Under the influence of environment, when the animal appears transparent, all the pigment is contained in the middle of the cell while the ramifications are free from pigment. When the animal appears brown both pigments are spread out into ramifications. In the condition of normal spreading the animals appear black, while many animals show transitory changes in color under influence of their surroundings; in a few cases permanent changes can be produced.

"The best examples of this are those observed by Paulton, in the chrysalids of various butterflies. These experiments are so well known that a short reference to them will be sufficient. Paulton found that in gilt or white surroundings the pupae became light colored and there was often an immense development of the golden spots so that in many cases the whole surface of the pupae glittered with an apparent metallic luster. So remarkable was the appearance that a physicist to whom I showed the chrysalids suggested that I had played a trick and had covered them with gold leaf. When black surroundings were used they were as a rule extremely dark."

Just notice the facts in these cases: first, a picture of the ground is taken in the eye, transferred to the skin cells through several stations or nerve centers,—they must first get a picture of the outside surroundings; second, the cells of the skin then arrange the color plugs in the surface of the skin in such manner as to effect the desired shade. The facts prove that it is first a vision transmission by a method of telephotography. If the cells can build instruments of that kind it is reasonable to think that man will also be able to produce the same kind of instrument. Not only will these fishes produce the same color on their backs as the bottom of the sea but will also produce the same appearances of stones, pebbles and other features to correspond with their immediate resting place.

Now this should illustrate fully the idea that it is not the fish, God, destiny, nature nor the devil that builds and instigates the actions of plants and animals. Here we have the cells arranging the pigment to conform to the surroundings, as soon as they get a picture of it, so as to have something to go by. Still Mr. Loeb believes, or at least claims, that the cells have nothing to do with it, that it is merely a chemical action, while the minister says God's will or Divine providence is the cause.

Let us consider some of the common emotions, like love, fear, and jealousy. Take for instance the actions arising from love. A person can be perfectly sane and still be under the control of those passions or emotions, and while possessed of them commit some of the most absurd and insane acts. The cells concerned with sex are given full control of the body. A picture of a certain female has been transferred through the eye to these cells and they have decided that the possession of that particular female must be obtained at any cost. A very singular but significant thing has been shown, that the actions of the male and female germ cells in the acts of copulation as single cells are very similar to the actions of the individuals, from whence they came. The fact is that our actions are directed by the sex cells and not by the brain cells. This fact has been fully conceded by other scientists, as will appear by the following from Ribot on the emotions:

"Solomon says that jealousy is as cruel as the grave. Shakespeare agrees with the Bible and holds that only fools are jealous. He also shows with wonderful insight that jealousy is self-destructive and its venom more deadly than a mad dog's tooth.

"Jealousy undoubtedly is a magnifier of trifles, and the most accurate and precise mathematician or bookkeeper correct to the letter ordinarily, will when jealous turn a simple sentence into volumes of pseudo-history. Trifles as light as air are to the jealous confirmations strong as proofs of Holy Writ. A jealous woman believes all that her emotions suggest. It is her jealous nature to swell small things to great. Out of nothing she will conjure much and then lose the truth amid the hideous phantoms she has formed. She makes a true husband false by forever suspecting him. Inquisitiveness never allays the pangs of jealousy. You may as well expect drinking in a fever to cure a thirst.

"Dr. Savage's first difficulty is the well known one of defining what he means by supposedly simple words. 'Jealousy,' he says, 'is a personal feeling where you feel powerfully a want of something which you are deprived of by another. It is a feeling and not a thought.'

"So intimately is jealousy usually associated with one of the other sex that you are prone to lose sight of the fact that dogs are often jealous of new babies, children of the same sex are jealous of each other, professional men dislike each other because there is an unconscious jealousy present, which they not only deny, but fail absolutely to recognize. Close competitors are not only jealous of each other, but particularly and more strongly jealous of those whom they see forging ahead of them. Jealousy always has in it a suppressed, hidden and irritating admiration or fear of the one you are jealous of, unless it is of the opposite sex. Then love or passion or the sense of possession comes also into play. When a medical authority dislikes another physician, he is unaware of the fact that he is jealous of the other man's prominence, progress and work and is equally unconscious of the fact that he himself feels a sense of dispossession. Therefore he hates him and envies him, although he is convinced that such a thing is beyond all possibility. Just as hunting animals are always jealous of others sharing their prey and will fight them, so a jealous woman will fight all other women with her handiest weapon, the tongue, and a jealous man will stop at no means to be rid of the source of his envy. Even fish at spawning time show fight to other males that dare approach their spawning mate. 'This,' says Sir George Savage, 'is important because the male fish and its mate are never otherwise than mentally in touch, for the female fish lays its eggs outside and they are fertilized there—not within the body.'

"In birds sexual jealousy occurs before as well as at the time of the period of mating. The struggle is often between parent and offspring as to possession of the female. In the case of mammals, the struggle is fiercely in evidence. The meekest of mice and the fiercest of wild beasts really seek out combats for the possession of their mates. This is also true of apes, ourang-outangs and other monkeys. . . . . .

"If the individual with this surplus of jealousy making

agencies is a self-depreciative, retiring, bashful person, suicide may result. If, on the other hand, he is of the bold, forward, belligerent, noisy kind, jealousy leads to homicide. In either instance, insanity is the motive force which has loosened the check reins of the will."

You see in this statement it is agreed that the brain may be perfectly sound, but the body is placed in control of some nerve center, or little brain, as they call it, located somewhere in the body and not in the brain.

Take another emotion, for instance, that of fear; no doubt certain special cells are given full charge of affairs, when a person becomes frightened. He does things swiftly and without thinking. Stampedes and other irrational acts take place, as for instance in fires and wrecks. A good illustration of an animal in charge of special nerve centers controlling the body in case of fear, is the frightened horse. When once frightened or possessed by the emotion of fear, he will run as long as he is able; he will stop at nothing. Certain motor nerve centers in control of the movements of running will compel him to run as far and fast as possible. On the plains where the horse lived for thousands of years in a natural state, his only method of escape from his enemies was a run for his life in this manner. The cells organized for that purpose keep it up even now when it is no longer of any use to the horse. His instinct to shy to one side of any object is evidently also located in certain cells. No doubt tigers and lions always hid behind different kinds of objects and the horse soon learned that it was the safest plan to steer to one side of all objects, no matter what they looked like. It is a singular thing how suddenly and completely the nerve centers in charge of the emotion of fear can take possession of an animal.

I have watched the actions of deer and elk. The deer

will be grazing quietly and if a dog or sometimes a man gets in the way of his olfactory nerves, he will perceive their presence and will instantly start off at full speed. The instant that the cells in his nose by a chemical analysis discover that the dust particles in the air are from a dog and not from a deer, the motor apparatus is given full control and he moves to safety immediately. If, however, the smell should come from another animal, like a cow or horse, the deer will raise his head and at first look and investigate, before it starts running. This is precisely what would take place in a similar case with a battleship or submarine.

Every cell of the deer's body knows that the dog is his enemy, so there is no need of stopping and investigating the matter, but upon discovery orders are immediately given to get away. So it is with the English ship; as soon as it discovers the submarine to be a German vessel, the only thing to do is to escape. If, however, some ship is in sight which has not the appearance of a German vessel, the only natural thing to do is to further investigate. It is not chance that controls the actions of the deer but an intellect whose unsleeping watchfulness and loyalty is equal in every way to that of man. It is the intellect of the cell that keeps careful watch over all that happens within and without the individual. In the same manner it is the intellect and wisdom of man that guides the destiny and welfare of the English ship.

A chicken raised from an incubator who has never heard or seen a hawk, can tell a hawk from a crow at a distance of half a mile. They will run to the house for safety at the first sight of a hawk, but pay no attention to other birds. In similar manner the crew of a battleship can tell at a distance the first time they see a ship, whether it is an enemy or a friend.

The experiences of the ages are in the intellects of the individuals that control and guide the battleship, and not in the battleship itself. So it is with the chicken that knows from past experience what kind of bird is dangerous. The chicken has no more to do with it than the battleship. The intellect that guides the ship will make mistakes now and then, and so will the intellect that guides the chicken.

A most singular fact observed in the life of the cell as in larger animals is that competition exists among them in the same manner as among the animals and plants they build. It is the intention of the cells that competition shall exist. Thousands of male cells start out to find the female cell at about the same time. The intention is evident that there shall be fair play and the best man shall have the opportunity to live and perpetuate his kind. We find the same arrangement among the Social bees. There is one Queen, and sometimes five hundred or more males. The male that is the swiftest and overtakes her in her nuptial flight will be selected to perpetuate himself.

It is a singular thing that in all organic life everything is made for a purpose, that is, every animal is built with a design or purpose to lead a certain kind of life. The muskrat is a structure made for the purpose of moving in water and on land. Its fur is adapted for water. It has a thick fine hair oiled so that the skin may be in water and still remain dry. Its feet are wide and flat, and webbed so that it can swim. Its tail is wide and flat covered with scales to assist in swimming. The house rat is different and is fitted for its purpose and environment. In the same manner every organ is made for a particular purpose or use, which we may call function. Every organ of a body, plant or animal is a body of cells grouped together for a certain purpose, and assigned a certain work to do.

The more highly organized any being becomes, the more specialized and complex are the organs and the work performed by them. In the lower beings like the polyp there are no cells assigned to the special work of receiving impressions and giving orders, and for that reason this plant or animal, whatever you wish to call it, has not what we call a nervous system: so if it is touched it will act and pull in its tentacles, but its actions are slow and from one tentacle to the other until they are all in; while in others just like it in appearance, but which have cells arranged for the special work of receiving sensations and giving orders, like the polyzoan, the tentacles will all be pulled in quickly and at the same time.

These animals have what we call a nervous system. These nerve cells have no other work to do and they become specially adapted for this specific work and are able to notify all the cells to act at the same time. In the same manner the organs of every being become adapted or in other words adapt themselves to the particular work assigned to them.

The heart is a combination of cells arranged into groups to form muscles, to push the blood in one continuous direction. The lung is a place where the cells come in contact with the air, and breath and carry with them enough oxygen for the stationary cells in the bone, muscles, nerves, etc., whose work prevents them from coming into direct contact with the air in the lungs. There is at all times waste matter accumulating in the blood, so the liver was organized,— which is another group of cells with the special work to do of removing waste matter from the blood. This is also true of the kidneys and the bladder.

There must be a mill to grind the food, before it goes into the stomach, so teeth are built up at the upper end of the oesophagus, which we call the mouth. Every thing is a group of cells, stationed at a certain place and assigned a certain work to do, each group performing its alloted work, faithfully and promptly, as is absolutely necessary for its existence.

If any organs like the heart, kidneys, liver or lungs should stop working, or even do the work poorly, all the others would suffer. So when the work of any one is interfered with the whole body is threatened.

It has been demonstrated that the nerve center which has the breathing apparatus in charge, observes constantly the condition of the blood flowing through it, as to its aerated condition. This fact was demonstrated by directing well oxidized blood through it, when the breathing would stop, and when poorly oxidized blood was directed through it the breathing became very rapid, regardless of the condition of the blood in the body in general.

We see from these facts that every movement of our body, voluntary or involuntary, is performed by reason of being instructed to do so by certain nerve cells or nerve centers, who have that particular work to attend to. We have seen that a nerve cell is a cell modified a little, and adapted for a particular work.

Any way that we look at this matter of evolution and development in life, we see that the actions of the individual must be the actions of the cells that occupy it. Darwin and several others argue that instinctive actions must have arisen at some time from conscious practise. That is, the conscious practise must have been kept up until it became a habit, and then the habit was inherited. Let us briefly examine that proposition. The young bee will proceed without any previous instructions to build the cells of the comb, just as perfectly as the older bee. If for any reason she gets started wrong, she will tear it all down and start it all over again. She will work at it until she gets it right. Where do these actions appear like habit and when did the bee ever have a chance to practise in a conscious way, in the past any more than today, so as to form a habit? You see the idea is absurd. Why should the bee's acts be instinctive any more than Mr. Darwin's? Why should the bee's actions be considered blind and unconscious?

Take another case of the plant that makes traps in which to catch insects. My school book on botany describes it as follows: "The leaves of pitcher plants form tubes or urns of various forms, which contain water; and to these insects are attracted and drowned. The common pitcher plant of the northern states, a Sarracenia, is a well known bog plant, but is not so elaborately constructed for capturing insects as is a common southern Sarracenia. In this plant the leaves are slender, hollow cones, and rise in a tuft from the swampy ground. The mouth of this conical urn is overarched and shaded by a hood, in which are translucent spots, like numerous small windows. Around the mouth of the urn are glands which secrete a sweet liquid, known as nectar. Inside, just below the rim of the urn, is a glazed zone, so smooth that insects cannot walk upon it. Below the glazed zone is another one, thickly set with stiff, downward pointing hairs; and below this is the liquid in the bottom of the urn. If a fly attracted to the nectar at the rim of the urn attempts to descend within the urn, it slips on the glazed zone and falls into the water; and if it attempts to escape by crawling, the downward pointing hairs prevent. If it seeks to fly from the rim, it naturally flies toward the translucent spots in the hood, since the direction or entrance is in the shadow; and pounding against the hood, falls back into the water and is drowned. Now you will have to say according to this theory, that instinct arose from conscious practise at some time in the past. When did the plant ever get the habit of eating insects by conscious practise, or when did the plant have a chance to get the habit of building this wonderful fly trap, by conscious practise?

If the actions of the bee or plant were conscious in the past, they must be so now. These absurd ideas arose from man's egotism. Man thinks his brain cells are smarter and different from the cells of any other animal, insect or plant, and the fact is, they are all alike, one as smart as the other. When the microscope first discovered the cell for us, we gave it the name protoplasm. The name is meaningless and misleading. The cells are the superior beings that create all living things and as such are the most highly organized beings in existence. All actions of plants and animals, habitual, instinctive, emotional, reflex or impulsive are caused by cells. Watch the actions of a squirrel or bird building its nest. He knows what he wants, where he wants it and how to build it. Still he has this knowledge without having had the slightest education or instruction in the work. His nest is an artistic production. He selects and carries the right material, weaves it together and, like any skillful workman, makes the forces of nature subserve to his wants. Like man the bird builds and weaves; like man the spider lays snares and prepares an abode for its young; like man, the squirrel gathers provisions for the winter; like man, the caterpillar makes itself a coat. In other words, the animals without any previous instructions practice all the industries of man.

We say that man's acts are intelligent because he does these things purposely and knowingly for an end. Animals, we say, act blindly, not knowing the purpose of their actions; except for the fact that man acts knowingly and the animals blindly or unknowingly the actions are the same, that is, intelligent. What right has man to make such silly statements, being himself an animal? Why should his actions be any more knowing than the actions of the other animals?

In regard to instinctive actions of insects and plants the following by Mr. Walker is interesting:

"In the pitcher plants the leaves take the form of upright pitchers. The upper secretes honey, which attracts numerous insects. Below this zone is a slippery surface, the slide-zone, which causes the insects to fall into the water below. These hairs are inclined downward, so that they catch and prevent the insect from climbing out; they are drowned in the water at the bottom of the pitcher, where they decompose and provide food for the plant.

"It has been said that a further adaptation takes place in one species of pitcher-plant (Nepenthe bicalcarata). The species occurs in Borneo, where a little lemur (Tarsius spectrum) has learned that it can get a number of insects from a pitcher-plant without the trouble of catching them for itself. This particular pitcher plant has got the better of the lemur, however, for it grows two long, strong prickles from the lower side of the base of the lid. These project downwards into the opening of the pitcher, so that the would-be robber is seriously scratched.

"The manner in which some other carnivorous plants catch their prey is equally remarkable. The leaves of Drosera are provided with numerous tentacles, at the ends of which are glandular enlargements. The tentacles round the margin of the leaf are long, those in the middle are short. The glands secrete a quantity of a glistening and very sticky substance, which insects mistake for dew or honey and are attracted. Now if the short central tentacles are stimulated, all the long peripheral tentacles bend over so that their enlarged ends cover the middle of the leaf. If an individual long tentacle is stimulated, only that particlar tentacle moves.

"Let us see what happens when an insect touches any of the tentacles of the leaf. If it touches the central tentacles, it is held by the sticky substance, and all the long peripheral tentacles bend over and enclose it. If it touches one of the long peripheral tentacles, it is also held fast by the sticky secretion, and the tentacle bends over and deposits it on the centre of the leaf. This stimulates the short central tentacles, and so all the other large tentacles bend over and enclose the victim. When the prey is thus secured, the glands proceed to secrete a ferment that digests the insect, and the products are absorbed by the leaf.

"In Scorpirurus the pod containing the seed in some species closely resembles a caterpillar, with the result that insectivorous birds are tempted to seize it. They probably carry it some distance before discovering the mistake, and thus distribute the seeds. In Acanthorhiza adventitious roots provide a thorny palisade above the ground at some distance from the tree, thus preventing the approach of animals. The extraordinary instincts observed among insects have been brought forward in support of the theory that acquired characters are transmitted. The instincts of Ammophila and Sphex, two genera of fossorial wasps, have been given as particular cases. The larvae of Ammophila feed upon caterpillars, those of Sphex upon crickets and grasshoppers which are supplied by the female. In both cases the female wasp first makes a nest, then drags the prey into it and lays her egg or eggs upon the animals which were destined to provide food for the larvae when they hatch out. The female then seals up the nest and does not visit it again. Now, in providing food supply for the future larvae, several extraordinary instincts come into play. It is necessary that the caterpillars, crickets, and grasshoppers should be paralyzed when they are caught, otherwise the wasp would have tremendous difficulty in conveying them to her nest, for frequently the prey is much larger than the wasp. Again, when the larvae are hatched out, they are very delicate in structure, and if these comparatively large animals were able to move about freely they would probably kill the larvae instead of providing them with food. On the other hand, if the prey were killed outright by the wasp when caught, decomposition would set in in a few hours, and thus the provision made by the female wasp would be useless to the larvae. This difficulty is, however, overcome by the wasp. When Ammophila catches a caterpillar she stings it in each of the segments of the body. In the caterpillar there is a separate nervous centre (ganglion) in each segment, and if only one or two were stung, the other segments would still be able to move quite freely. In the prey of Sphex, however,—grasshoppers and crickets—there are three separate nervous centers (ganglia) which control the movements of the animal. These are situated in the thorax, and when the Sphex catches her prey she stings it in these three separate ganglia. The result of this is that the prey is rendered unable to move, but is not killed. Fabre, who first described these phenomena, was at first unable to understand how it was that the prey he found in the nests of these fossorial wasps did not decay, but he subsequently showed exactly what happened by some very ingenious experiments.

"He took some of the crickets and introduced poison into their bodies with a needle. When this was done indiscriminately, that is, in no particular part of the animal's body, it either killed the animal or produced very temporary results, according to the amount of poison introduced. When, however, he introduced the needle into the three nervous centres (thoracic ganglia) which control the movements of the insect, he found that complete paralysis ensued. More than this, the metabolism was checked, and the paralyzed insect continued to live without food for several weeks, which it certainly could not have done under normal conditions. An interesting point about this instinct of Sphex is that the prey is stung in one particular point where the tissues are soft and the nerve centres are easily reached from the surface. Much the same thing happens in the case of Ammophila and its prey, the caterpillars, only here the wasp has to apply its sting many times, so that all the middle segments, at least, of the animal's body are paralyzed.

"Now the supporters of the theory that acquired characters are transmitted say that it is impossible to account for the origin of these instincts in any other way than that the ancestors of the existing wasps first exercised a certain amount of intelligence. A wasp found that when it stung a grasshopper in a particular place, that grasshopper was paralyzed and was much more easily carried to the nest. It remembered this, and led by past experiences, always stung its prey in the same place. This habit produced an effect on the germplasm, and the offspring tended to sting their prey in the same manner until the instinct became established, and so no longer depended upon the intelligent action of the wasp. This interpretation, however, appears to break down when carefully considered. To begin with, it assumes that intelligent action preceded instincts. We find that the higher we go in the animal kingdom the fewer the instincts and the greater the intelligence. We only find intelligent action as a very late product in evolution.

"To place intelligent action before an instinct, in fact, to derive instincts from intelligent actions, seems to be placing the cart before the horse. Another very important point is that the capture of the grasshopper, crickets, and caterpillars, paralyzing them without killing them, and storing them in the nest, is not of the slightest use to the individual wasp. It simply provides food for the larvae which the wasp will never see, and of whose needs it can have no experience. This interpretation, in fact, attributes to the wasp a prophetic knowledge with regard to subsequent events, of which neither it nor any of its ancestors have had any experience."

You notice Mr. Walker has considered the idea of intelligence preceding instinct, and discovered its absurdity. You notice the wasp is a skilled hunter and surgeon and a good provider. He provides for the future offspring which he will never see, just as intelligent man provides for future offspring, which he may never see. What right has man, who is made and guided by the same being, the cell, to say that these same acts in the wasp are less intelligent than his. The animals that the wasp captures and carries home to his family are three or four times larger than himself. He has by practise through generations and ages learned that the caterpillar has certain cells that direct the actions of all other cells. These are called nerve centres. He has also found that if he destroyed these the other cells would still live but would not be able to direct any movements of the body of the caterpillar. After having made that discovery, he saw the advantage of paralyzing the caterpillar in that way, so that his young would be provided with fresh food. But, you will say, how can the wasp know or learn this method of partly killing the caterpillar, when he has never seen it done, or ever received any instructions. There are only a few microscopic spots that must be punctured, while there are a million other places to puncture, where it would not be effective. How can he know the right spot, not having ever seen a caterpillar before? The answer is very simple. The builders of this flying structure, the wasp, have provided it with tools, and to think that they would not know how to use them would be a foolish idea. The cells that built the wasp had the experience of ages fixed in their memory. They had been practising the dagger thrusts while in other wasps before him, for centuries. Every movement and thrust of the dagger on previous occasions would be fresh in their memory. The wasp is but a flying machine with tools to affect these several actions. The cells have made the machine and the tools and they know how to use them.

Let us now consider the pitcher-plant of Borneo, that has trouble with a species of ape called the lemur who steals the insects from his trap. This plant, who is of an inventive turn of mind anyway, soon discovered or conceived some scheme to prevent the ape's stealing the insects from its trap. The fly trap itself is a wonderful invention. It is a most perfect and scientific arrangement to effect a certain purpose; however, you know that the plant itself could not build these things, any more than a house could build an addition to itself.

The builders and occupants must be given credit for the means and the intelligence evidenced by these inventions and productions. What difference is there between the mechanical skill required to build a heart with which to pump or push the blood through the system, and that required to build the fly-trap. It was a bright idea of the plant to provide these prickles, in order to stop the ape from stealing his property. It is, however, equally intelligent to invent the eye, more so than the specs and telescope invented by man to aid his instrument in seeing smaller or larger objects for which it was not made. The cell builds instruments we call ears, with which to catch the vibrations of the air; man builds ear trumpets to aid him in hearing better. All these are instruments and productions of art, and can only be produced by a mind or intellect. The cause must be mind or intellect, which can direct and mould the blind forces of nature into those things desired by the builder.

The purpose of this book is not to explain the origin of the cell and its high organization and what the mind and intellect in the cell really are. In this discussion it makes no difference what you call this quality or ability to mould and direct crude matter and blind natural forces. Call it soul, vital force, nature, God or intellect, it makes no difference, for it is in any case the cause of all living constructions. In man it is called intellect; so we may as well call it that. As in the cell, it is the intellect in man that is the cause of his structures, like houses, railroads and ships. What difference can there be in the acts of the cell making a crystalline lens for the eye and the act of man making lenses? What difference can there be in the act of the cells in the chicken gizzard, using stones to grind grain, and the analogous act of man grinding grain with mill stones? The instruments used and the purpose are the same.

Is there any difference between the industry of the cell

which makes the body or organs or instruments of the body, and the human industry which makes a building and machines, similar but on a larger scale? We have called the intelligence of animals instinct, but, as I have stated before, a number of scientists of late have admitted that the instinctive acts of animals are intelligent acts.

The cell that directs the involuntary actions of organs inside of the body sometimes directs the voluntary acts outside of the body. In the work of the cell as well as in the work of man, there is a relation of the parts to the whole and also a method by which the part or the whole is to be used in a machine or structure made by man, as well as in a machine or structure made by the cell. Like a plant or animal, each part has a meaning or place in the general idea or scheme of the whole machine or structure. Every part has its place in the whole machine.

The periscope is an instrument with which the submarine can see. We may liken it to the human eye. These are, however, mere instruments back of the periscope, while back of the eyes are the individuals whose intellect must judge and decide and direct the actions of the individual, be it man or submarine. In each case the purpose of the instrument is the same. The eye could see nothing without the cells in the brain to receive and interpret the pictures transmitted. Nor could the periscope see anything without man, its builder, who is there to receive and interpret. All the organs of our body are instruments put together like a machine for a purpose, each part made to fit in and work in harmony with every other part. You see the cell, like man, is a skilled artist that pre-arranges each and every part of the whole. We see with our eyes how the cell attends to all the details of maintaining the body in proper repair. We know that to do so requires about the same skill and intellect as to build it. The skilled engineer that is able to run and keep his locomotive in proper repair is considered just as intelligent as the builder. His intellect must comprehend and understand the relation of each part of the machine to the whole. A sure and certain proof that cells are able to judge, remember, reason and direct, is that they do so in our head.

It is not the man who reasons, it is the cells occupied with that business in his head. If a wound in the arm destroys a main artery carrying blood to the hand, a new artery will be constructed around the wound, of a sufficient size to carry the necessary and usual amount of blood to the hand. How can natural selection have anything to do with these new actions? The ability to do this requires the best skill and judgment. Who determines the circumference and size of that artery? In these acts there must be some one who decided on what should be done in each particular case. The power to do this can be accounted for, by their general knowledge of arteries gained from ages of experience, and also the experience had in the construction of the body to start with, and in its maintenance. From this experience we must presume they know how to make a new artery at any time and in any place. The builders of any structure would necessarily possess the knowledge of how to repair any particular part of it. It would not be as good as the original structure, and would not look as well, but it might serve the purpose for which it was intended for a long time. It is just the same with repaired organisms. Nerves are strings or bundles of cells who have a certain line of work to perform, in receiving impressions and giving orders. By use they also become expert, as we see in the nerve cells of the fingers of the blind, who can read by feeling. In the same manner muscles are strings and bundles of cells that have only a special work to do and by use they become expert and stronger.

This ability to adapt itself to what is required in any particular case that may arise, shows that there is some being ready to do the right thing at the right time in any and all emergencies. It has the ability to do a thing that never was done before. I have an illustration of that now on my farm. We have some Angora goats; three of them have walked on their knees for a couple of years. I examined their knees a few times and found that a callous substance, just like hoofs, has formed on their knees. They get diseased feet from walking on the soft wet ground, and for that reason walk on their knees. In order to continue walking on their knees, something must be provided for the knees to prevent the wearing through of the skin, flesh and knee cap. In order to prevent this, hoofs are provided in a place where they never were before. In this case and all similar cases, there is no chance for natural selection to get in its work nor have anything to do with it.

I can remember one of my tenants who had a pig one summer with a broken back or paralyzed hind legs, who propelled himself over the ground by dragging his body by his fore legs. This pig developed wonderful power in his limbs, and also developed a horny scale-like skin on the center part of his ham where it was in continuous rubbing contact with the earth, as it was dragged along. He lived and grew just like the others. This power to adjust itself to conditions is the power to change or construct parts different from whatever had been before. This shows clearly that the cells conceive ideas of what should be done, and some ideas are better than others, depending on what may happen to be required in the particular circumstances. In addition to this we find them trying out new and further experiments.

The cells of the body of any animal adopt such colors as are most beneficial to it, under the particular circumstances and environments which it has to contend with. If no color is necessary, none is provided, as we can see in the white skin of the human race in the North. This matter of deciding color, length of hair, or repairing broken parts, is an intellectual quality, as it will always depend on the intellect of the party who does it, whether or not it will come up to the requirements. We find that those animals and races of men with the best mental faculties will invent and try new and better ideas; they will always be equal to any emergency, when any contingency arises that must be provided for. The men with the best intellect will be the most likely to be able to do the right thing at the right time.

The cause of all living things is the intellect of the cell. Upon this fact we can proceed to discover other facts which may some day disclose the cause of intellect in the cell. We may some day by a more powerful microscope be able to see and understand all the different organs and parts of the cells, may be able to even see the actions of the primordial cell or being that organizes and builds the cells. But before the unknown, we can only guess.

A writer who believes that some God makes the bee, states, "Hardly are all the parts of the young bee dry, when it knows all it will have to do during the rest of its life. Let us not be astonished that it is so well instructed. It had been so by Him who formed it."

Now you see this expression is literally true. The bee can only get his instructions from those who formed him. As the entire scientific world now knows that the cell forms the bee, it should clear up the whole mystery of instinctive actions. Man that forms the ship and submarine must also instruct the submarine. It could not be otherwise. If the builder knows how to produce machines for a certain purpose, he will also know how to use them and instruct others how to use them. The plant which catches insects makes a gastric juice which digests and dissolves them. These actions are not different from those of man catching animals and preparing them. The object and purpose of the actions are the same.

The plant cells that make seed pods which look like worms or caterpillars, in order to fool the birds into carrying them around and scattering the seeds, show the same intelligence as man in his attempts to fool other people. The plant that builds fortifications some distance from itself, of thorns, to keep animals away, does so with an intelligent purpose, just as man has. However, in the case of plants, it is the cells that build the plant and furnish the intelligence.

The books are full of attempts to distinguish between instinct, reflex action and reason, but they all fail. The fact is, they are all intelligent acts. There is no distinction. The old popular phraseology, that animals act only from instinct and that man alone acts from reason, is absurd and without foundation.

The cell builds all animals and plants and directs the action of all. They are all intelligent acts. The insectivorous plants decoy and catch insects in large numbers. The single cell pursues and captures other cells and devours them. The actions are deliberate and purposive, just as in insects, animals and man; there is no distinction. The young swallow can fly in its first attempt because in the memory of the cells that made him and occupy and guide his body, there is stored all the experiences of past generations together with those recently acquired in the parents from which it came. The builders of the machine know how to operate it. There can be no more mystery about the cause of instinctive action than there is about the cause of a man's running a boat or a machine which he has built.

CHAPTER IX.

CAUSE OF EVOLUTION.

Evolution as understood by some is the cause of our existence. Evolution, however, means progress, growth and development. By others it is understood to be a cause of development from lower organisms to a higher. We can trace the evolution and progress of man gradually from the savage through the ages of stone, bronze and steel. With every new discovery he was able to better his condition and became more able to defend himself against the elements and his enemies.

The greatest discovery in the history of the world, as far as man is concerned, was the discovery of the club. From that instant man as he is now known and distinguished from other animals, was produced. It was the creation and birth of man. Until that moment, man had been at the mercy of the tiger, the wolf, and the bear and other animals. At that moment evolution or whatever you wish to call it took the crown from the lion and placed it on man. At that moment evolution as the judge proclaimed man king of the beasts. From that time man as an animal began to develop into a man. In order to be able to wield the club effectually he had to stand upright and he soon got the habit of walking altogether on his hind legs, from pure necessity of having to carry a weapon so as to be ready to defend himself at any mo-

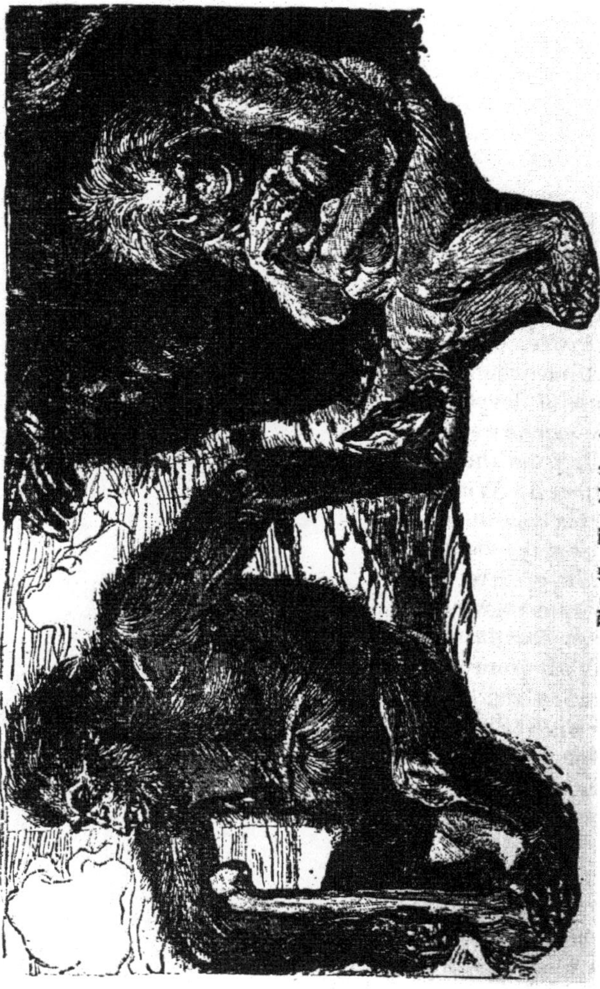

Fig. 45.—The birth of man.

ment. The crust of the earth discloses to us a continuous record of his progress since in the use of implements from those made of wood, bone, clay, stone and bronze, up to the steel and electricity of today. This natural progress is called evolution, but you will notice that the cause of discovery and invention by which evolution and progress are possible must be through the intellect of someone. The cause of the first act of discovering or defending himself with a club could not have been an accident or chance. The cause was the intellect and mind of the individual who conceived the idea. The use of a club in self-defense by primeval man in a fight with his enemies required the same will and skill as is required by him today in defending himself either with a club or with the larger weapons.

The club is a wonderfully effective weapon if skillfully handled, as one blow on the head will put man, lion, bear or tiger out of commission. However, it would have to be used with speed and intelligence. The idea of how and when and where to use it would have to be fully conceived in the mind of the individual before any effective use of it could be made. The primary cause of the act was and must have been the idea or intellect. The next thing necessary was the means with which to execute the idea. This individual was happily in possession of the means, the instruments with which to grasp and swing the club. The hands of the ape family are wonderful instruments for grasping and for that reason he was in possession of the instruments with which to grasp and handle a club. In his case all that was necessary was the idea. As soon as the mind cells in his head conceived the idea, it was then able to direct the actions of all the other millions of muscle cells who would have to do the work of using the club.

The legs of other animals are not provided with such grasping apparatus and for that reason, even if they should have conceived the idea of defending themselves with a club, the idea could not have been executed. The cell minds had produced these hands for the purpose of grasping limbs of trees in order to escape from other animals, but the hands of the ape and monkey tribe, or tree climbers, turned out to be very handy instruments with which to perform all kinds of work. Since the discovery of the club one idea has led to another and countless inventions have been conceived by the mind of man, leading up to such acts as sewing, weaving, writing, painting, etc. We know that today all inventions and discoveries must come about by reason of the idea first conceived by the mind, and no actions of any kind directed towards a purpose can take place unless first conceived by the mind of someone. Therefore, it seems clear that the cause of evolution, progress, and development is intellect in the cells who are the real directors of all acts, and not variation, environment, natural selection or survival of the fittest. These are mere phrases and combinations of words describing mere incidents and conditions that take place in the evolution of life. Take for instance beneficial variations. It is claimed that there are no two individuals just alike, that they all vary in some respect; that the one that by chance happens to be born with some variation that is beneficial to him will win out in the struggle for existence and be able to perpetuate his kind. Now this is generally true, but it is simply an incident. It does not build the individuals that do the struggling nor does it produce the variations. For instance, the gorilla is provided with a callous covering on the knuckles of his hands. It is argued that this callous covering is a beneficial variation to him because when he walks over the ground he walks partly on his knuckles and any gorilla that was not provided with calloused knuckles would not be able to get over the ground by aiding himself in this manner with his hands; so in this way, those gorillas with a callous covering over the knuckles would have an advantage over those without such a covering and would be preserved; in that way the callous was produced on the knuckles of the gorilla. Now we know after a full investigation that this kind of reasoning is without foundation because the cells that build the gorilla will provide this callous covering at any time, if necessary, and will also provide it at almost any place on the body. If the gorilla persists in rubbing his knuckles on the ground, they will have to be covered with something or else they will be destroyed and for that reason the callous develops on his knuckles.

Now while it is true that these beneficial variations like callous on the hands and knuckles might under certain circumstances be the means or cause of saving the life of the individual, still it must be clear that it does not in any way explain the cause of the callous, nor does the ground or environment produce these beneficial variations, as some evolutionists would have it. The cells are builders; they are expert chemists, artists, sculptors and mathematicians and produce everything that we use for food and clothes and a great number of things that we use for building material. They produce all our starches, sugar, acids, pigments, perfumes, wood, cork, fibers and so on, things too numerous to mention. If the knuckles have to be rubbed against the ground, they will have to be provided with a covering of such strength and durability as will stand the wear and tear arising from a continual rubbing contact with the ground. The ordinary hair and skin will not last long. Something tough and hard must be provided. We do not know how or from what the cells are able to build this horny, callous protection over the skin when necessary, but we know they do it. The evolutionists would make us believe that the callous is caused by the ground or by accidental chance variations, but it certainly is very clear that nothing of the kind has ever happened or can happen. The callous is a scientific production for a purpose, produced at the right time and at the right place. It was not produced until wanted nor was it produced on a place where it was not necessary.

The chemical knowledge and skill required to produce the callous show conscious intelligence of a high order. Take the simplest single cell like the foraminifer, which makes for itself a shell of limestone in which it lives. It leads a single independent life in the ocean and also associates in colonies. This animal is sometimes referred to as merely living matter. However, when this little animal discovered the secret of how to cover its naked body with a hard shell, it had the best of those other single cells in the struggle for existence and was able to exist in deeper water and resist the attacks of other cells. It was a great discovery and the chalk cliffs of England stand today as a perpetual monument to their effort and intellect. It required just as much intellect to select the microscopic lime from the water and mix it with the proper ingredients to make a strong shell and then form it into the perfect and proper shape, as it does to mix sand and cement into a mixture for the foundation of a house. Consider again the skill of the cells that can make bone or horns and can cut and destroy bone and horn whenever it is necessary; for instance in the case of a broken bone, the splinters will always be liquified and carried away. Even ivory pegs, if driven into bone, will be torn down and carried away by the cell. Children's teeth in the same manner are taken apart as microscopic particles and carried off, so that new ones can be formed in their places as permanent teeth.

Take another instance of the wonderful architectural skill of the cell in building the antlers of the deer and elk. The antlers are built up in the spring and summer and fall off again in the fall after they have been used and are not necessary any more for the purpose for which they were built. They are built for only one purpose and that is as a weapon with which to fight other male deer. While these antlers are growing, they are covered with a delicate skin called velvet, and through this velvet the blood circulates, carrying the multitude of working cells and the building material of which the antlers are made. The cells work together and build up the snags, beams and tynes that make up the antler. Like a hive of a million busy bees or skilled workers, they work beneath the warm velvet all spring and summer to form those enormous weapons, sometimes five or six feet high. In the early fall the cells quit work and retire into the body. The velvet under which they had worked falls like the autumn leaves and the hard bony weapons are exposed ready for battle. As soon as the antlers are ready the male deer is also ready to challenge another male deer to mortal combat and they fight for the possession of the female, who stands by ready to take the victor as her mate. As soon as the loves and battles are over and the mating is completed, the antlers are no longer of any use and they are shed. That is to say, they are cut off near the skull by those who formed them—the cells.

The deer is a nation of a billion or more individual cells who work together in harmony for one purpose, for one idea, the welfare of all, just as does the German or the

English Empire, whose purpose is to perpetuate and maintain its existence on this planet. To effect that purpose these antlers or fighting machines must be made. What difference can you possibly perceive in the intellectual capacity or foresight in the organizing, cooperating and building of implements of war by the nations of the world and the same act by the cell nations we call the deer and elk. Think of the gathering, grinding and mixing of the minerals required to produce the material with which to build the antlers on the deer! Think of the intellect required to guide and direct the details of the actions of those millions of workers! All this work of building the antlers goes on without any trouble or bother to the cells connected with the senses or consciousness of the deer, as they must be occupied with other and more important work, such as receiving information as to the approach of enemies and the place to find food and building material for the workers inside.

The white cells of our bodies, who are called the general inspectors and move through our bodies from one place to the other looking for enemies that may have got into the body in some way, such as bacteria, which they kill and destroy, show the same watchful care and loyalty as any human beings. The million of workers, who built up the antlers with only one purpose in view and then dropped them as soon as the fight was over, show the same intellect as that which guides and directs the actions of the German Emperor.

Chemistry and mechanics must be understood in order to produce the antlers. The chemist must first know what the effects will be of mixing certain substances in his crucible and applying the heat before he can go ahead and produce what he wants. Next he must know what he wants before he begins to build, if not, he could not possibly produce an instrument for any particular purpose. You may employ a man to build a house or machine and you furnish him with lumber, iron, brass, cement, clay and tools, but if he has not the knowledge to build what is wanted, if he has not a mental picture of the structure and a record of details of the work stored away in his memory, he can produce nothing. The cells that build the antlers likewise must understand their business. Men gather the iron, wood and other material to build cannons; the cells gather lime, minerals and other materials to build the antlers. Both work for a purpose and the constructive skill required in the case of one is exactly as necessary in the other. If man is an intelligent being the cell must be, too. A brain is not necessary to intelligence. The star fish, polyps and several others have no brains, still when their actions are examined, they show just as much intelligence in their place in life as any other animal. Their actions, however, cannot be directed with the speed of those who have a brain, or a common center to direct the actions of all.

A species of single cell that lives in the sea called deflugia, picks up microscopic grains of sand from the bottom, cements the grains together and in this way builds around himself a hard covering or armor we call a shell. In the armor the creature makes holes through which he sticks out his hands or feet and paddles himself through the water in search for food. This cell, as far as we are able to discover and understand the matter, has neither brain nor a nervous system, but still has a mind and intellect that directs his actions, similar to ours. Another kind of cell called the Arcella builds for itself a covering of a different material; and the method of building this coat or shell, which resembles in texture the coat or wing covers of insects, is not understood by man's limited intelligence. Another species called diatones build around themselves a covering like transparent flint, and are able to produce styles and patterns of great complexity and beauty. They imitate the designs produced by the natural forces in matter called crystallization. Man also does the same thing. In what manner do the actions of the cell building itself a protecting coat and going in search of food, differ from those of man when he also provides himself with clothes and goes in search of food. Another species of single cell living in the ocean has not only provided himself with a covering hard as flint and transparent as glass, but also with search lights. It is called the noctiluca, Fig. 46. Mr. Schute in his book on Evolution describes it as follows:

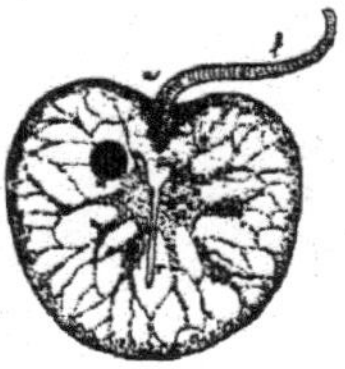

FIG. 46.—*Noctiluca miliaris.* Dorsal view.—SCHUTE.

"This little one-celled animal has the power, through its special chemical activities, of manufacturing and making light. It is through the agency of myriads of these little creatures that the diffuse luminosity of some seas is produced and can be observed at night. If the jar be placed in the dark and agitated in the slightest degree, there is an instantaneous display of light which is of a beautiful greenish tint and is so vivid that it can be observed in ordinary lamplight. This phosphorescence is only of one instant's duration and a short rest is necessary for each renewal." The special locality of the searchlights

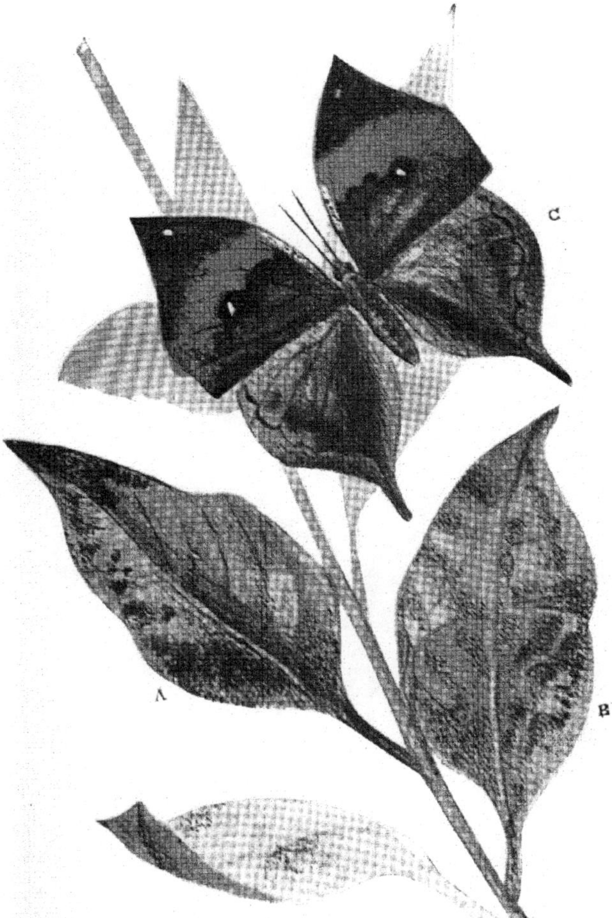

Fig. 47.—*A,* Butterfly of Sumatra, illustrating the wonderful skill of the cell in producing a work of art. *C,* Butterfly with expanded wings (dorsal surface), which are conspicuously colored to attract the opposite sex. *B,* Same butterfly with wings closed (ventral surface) and presenting a close resemblance to a dead leaf. *A,* in both color and shape, with the purpose of deceiving and escaping from his enemies.

has been found to be in the outside transparent coat of flint covering the body of the cell. Can you say that a being who can provide himself with an armor of transparent flint studded with flash lights, does not possesss intelligence, just as man does? The same covering could also be produced by the most highly civilized man for his use and defense, but not any better.

Coming back again now to the theories of the evolutionist, such as natural selection and the survival of the fittest producing the different structures in life, we shall again consider some of them. They say that any accidental and beneficial variation will help the animal in his struggle for existence and the variation will be transmitted by inheritance, as the creature with such beneficial variation will have an advantage over his competitor in the struggle for existence and will be able to live and perpetuate his kind. But the trouble about this theory is that the variations would not be of any benefit until fully developed; on the contrary, they would be a burden and a hindrance to the individual in his struggle for existence. Take as an illustration the pockets of the pocket gopher. The gopher uses the pockets to carry grass to its store house in the ground. He has a big pocket on each side of his head and these pockets he stuffs full of leaves and grass. In this way he can transport clean and fresh food to his underground warehouse with great ease and convenience and without dragging his food over wet and muddy ground; but until these pockets were completely formed, they could not have been of any benefit to the animal. It must be perfectly clear that these pockets on the pocket gopher could never have been formed by slight accidental variation. Take another illustration, of the gizzards of birds living on fish and meat; we find that while so doing, they have practically no gizzards and eat no gravel with which to grind their food, but if you change their diet to hard food, which requires grinding, they will also eat gravel and the stomach will gradually change into the ordinary gizzard.

What do all these things show? Simply this,—that the individuals in charge have not only sense enough to produce a particular kind of machine, but when it is necessary they know how to change it in such manner as circumstances may require and that nothing takes place by chance. It is very likely that some day man will be a toothless and hairless animal. Things are tending that way very rapidly.

The following from my daily paper I believe tells the truth: "The reason man has so much trouble with his teeth is that nature has concluded teeth are an extravagance for human beings and she is taking them away from us. This is the theory of Dr. Lloyd Marix, a London physician, who believes that man will eventually become as toothless as a baby and be glad of it. When man cracked nuts and bones and gnawed roots, to say nothing of his enemy's jugular vein, he had little or no tooth trouble. His jaw was, at least, a third bigger and protruded beyond his chin and even his nose. The teeth themselves were big, hard and more numerous. But nature has given man of today a bigger body, and what is still more costly of energy—a bigger brain. She has economized where she could, taking toll from the hair, teeth, nails, appendix, of anything that could be spared. A hairless, toothless race without toe or finger nails will be the final result, according to Dr. Marix. The gums will probably increase in height and act as a brace to the lips and be available in the way of substitutes for teeth, as organs of speech. Absence of teeth should be a blessing because by that time man is certain to have his chewing, and doubtless at least part of his digesting done for him by machinery. Dentists, barbers and manicurists will turn to more productive forms of labor preceded probably by soldiers, lawyers, and others who make their living on human misery."

It is only what we should expect when we remember that an animal is a nation or colony of still smaller individuals. If we begin to live on food that needs no grinding, teeth will be useless and will be gradually discarded. We, as a nation or as individuals, would do the same thing. Nothing will be produced and maintained except what is necessary and for a purpose. Take for instance the case of bees and ants. They are not as large as man, still they display the same intelligence in reference to their affairs as man. Their social actions and organizations are just like those of man and the cells that build man. The history of the gradual social evolution of the bee is the same as that of man. The following from a newspaper is a good description of it:

"The bees are like human beings in this: they get the habit of work and sometimes they continue working for a little while even after work becomes unnecessary—as you see a rich old man still working, though he need not work, but bees, like others, are spoiled by prosperity, and before long all of them lose the habit of working.

Buchner, the German scientist, observed that near sugar factories in the Barbadoes, the bees give up the troublesome work of visiting flowers. They soon learn that they can get all the sugar they want all the year round—which is what the sons of rich men learn very easily. In warm countries, where flowers bloom all the year, the bees give up the storing up of honey for the winter—they forget about the cold and rainy days and only gather enough honey and pollen for each day. To make such bees work, it is necessary systematically to take from them what they have gathered. Like the workers in many tropical countries, who disappear when they have enough money to last them for a day, the tropical bee loses the habit of labor.

There are almost 5,000 varieties of wild bees and among them patient study would undoubtedly discover peculiarities as numerous as those among the different kinds of human beings. You may even find among the bees primitive, ignorant, uncivilized individuals, corresponding with the Bushmen of the desert or even with the cave man of a hundred thousand years ago. One little wild bee called the Prosopis you may see flying about in the bushes. If you knew bees and could study these they would seem to you as different from the prosperous bee of the hive as a half naked bush-woman seems different from the comfortable lady in her furs. But that little wild bee, half starved and ignorant, is the ancestor of all the civilized bees. And what is more important, as the scientists point out, it is probably to her that we owe nearly all our flowers and fruit. A hundred thousand varieties of plants would disappear from earth if the bees did not visit them, carrying the pollen.

"We human beings would understand ourselves better if we knew more about the insects that live in organized civilization at our feet or in the air above us. In proportion to their power the bees and ants are infinitely more highly civilized than we are. They are as far above us as the careful, painstaking worker and saver is above the worthless tramp and idler. For thousands upon thousands of years before men had dreamed of civilization or settled in great communities, the bees and ants were working out their problems of co-operation, organization, defense and attack. Aristotle wrote about the bees and their civilization, their laws and their habits, more than 2000 years ago. With scientific exactness and from his day to our day in which such men as Darwin and Buchner have studied the planning, lawgiving insect kingdom, there has been in all probability no change whatever in their lives. Long before man had cities or villages, the bees and ants had their great nation. Thousands of years ago they knew all that they know now and they probably know now all that they ever can know. So, at least, it seems to us bigger creatures who study them. They are our superiors in proportion to their strength, but we are still developing, and therein lies our hope."

What is here said of the bees and ants, of their high state of civilization, organization and co-operation, can also be said of the cell, who builds the bees and ants. For thousands and thousands of years before the bees and ants and man had dreamed of civilization or settled in great communities, the cells were working out their problems of co-operation, organization, defense and attack. We can trace the gradual social evolution and progress in the cell, the bee and man. I do not see that the size of the things produced can have anything to do with the intelligence required to produce them. It is just as difficult to design and make a good watch as it is to plan and build a threshing machine. It must be just as difficult to design and arrange a structure made of atoms, molecules or grains of sand as one made of bars of iron, stone or brick. It is now well known that the cells lining the stomach and intestines stick out their hands and grab those particles of food and building material that are required. They select and choose what is wanted and leave the other alone. As the food and other material is selected, it is handed on to the lymph cells, who carry it to the places where it is needed. Mr. Binet has the following to say about this:

"To illustrate, it was at one time conceded that the phenomena of resorption and nutrition were explainable by diffusion and endosmosis; Dutrochet, upon his discovery of endosmosis imagined even that he had discovered the principle of life. At the present time we know that the walls of the intestines do not in any wise act like the inanimate membrane used in experiments in endosmosis. They are covered with epithelial cells, each of which is an organism endowed with complex properties. The protoplasm of these cells lays hold of the food by an act of prehension exactly as the ciliate Infusoria and other unicellular organisms do, that lead an independent life. In the intestines of cold-blooded animals, the cells emit prolongations, which seize the minute drops of fatty matter and carry them into the protoplasm of the cell, convey them thence into the chylifactive ducts. There is still another mode of absorption of fatty matter, met with among cold-blooded as well as warm-blooded animals. The lymphatic cells pass out from the adenoid tissue, which contains them, so that upon arriving at the surface of the intestines, they sieze the particles of fatty matter there present and, laden with their prey, make their way back to the lymphatics."

It requires just as much skill and intelligence to select the required material from the stomach as ordered by the cells in other parts of your body as it does to give the orders for the material. The cells in the stomach must be just as intelligent as the cells in your spine or head that give the orders for material with which to build hair or the callous in your hand. It is clear that when the cells begin building the plant or animal from which they came they must have in their mind ideas or a mental picture of what they are about to do. Not only must they have a mental picture of the particular structure that they start out to build, but they must also have a mental picture or knowledge stored in their memory of how they shall go about to accomplish the work.

It is, however, nothing more than reasonable to expect that they should know how to do this, as they have performed the work millions of times before. We see in our body that the white cells are soldiers, which have not been stationed at any certain permanent work as nerve cells, muscles, etc., have, but have a general knowledge of performing almost any act that may come up and may be required as necessary to the welfare and existence of the body. In the case of the healing of a fracture of a bone or a wound, some will take the job of building epithelial cells; some will build connective tissues and some will build bone; others will destroy themselves in their effort to destroy invading hordes of dangerous enemies like germs and bacteria. They all go to work immediately in case of a bruise, cut or broken bone with the skill and knowledge accumulated in the past ages in their memory, each at his job, knowing just how to do any or all of the work, each and all willing to do anything that may be necessary to affect a repair, each one working in harmony with the other in such manner as not to interfere with the work of the other. Thus alone could it be possible for them to fix up a broken leg or other bruised, crushed or mutilated organ.

The religion of the cell is similar to that of the Japanese. Patriotism is the expression of his religion. The mind of the cell does not split hairs as to his individual rights or the rights of the body or of the race. So far as he is concerned the welfare of the body and its perpetuation is the whole object of life. The body is the object to live for and to die for. The cell is not an individualist. He has developed both moral and national consciousness. He is not interested in his own welfare, except in so far as it is the welfare of the body. He is in a similar relation to nature as the bee worker is to the hive, himself nothing, the body or cell community everything.

I repeat, the religion of the cell is similar to that of the Japanese and the most admired quality of the Japanese is his patriotism. Mr. Huxley describes the sponge as a large city and states: "The sponge represents a kind of sub-aqueous city where the people are arranged about the streets and roads in such manner that each can easily appropriate its food from the water as it passes along."

As a matter of fact, every animal or plant is a city of some sort, sub-aqueous, aerial or terrestrial and occupied by the cells who have built it. Every such city is built on a plan or with a design or purpose to take care of the millions of inhabitants who occupy and maintain it. These cities or abodes when examined are found to be constructed with a purpose in view or a design to meet a condition or existence of some particular kind. They are found to be constructed with wonderful skill and design, to meet the most severe and complicated conditions in life.

Upon close investigation it is found that the cells who build the city we call the sponge are just as intelligent as the cells that build the fish, animal or man. They understand how to build the house on the bottom of the sea of such material, lime and fiber, that the other animals cannot make any use of it and so will not eat them. They understand how to cause a continuous stream of water to flow through their protected habitation, and in this stream of water, they pick up their food and other building material they may need. The young sponge starts out in life as a single cell and swims around in the ocean for months until it finds a mate, then they settle down and commence housekeeping on a suitable rocky place on the bottom of the sea, and there they begin to multiply in numbers and build an enormous colony which we call the sponge. The building process, however, is quite slow, as they gather the food and building material from the sea water.

The sponge building cells live on smaller cells usually called bacteria and microbes. These smaller cells are by no means less intelligent in their place in life than the larger ones. The cells that know how to make their own food like starch, sugar and other carbohydrates are generally smaller than the other cells. They usually make only stationary abodes like plants and trees. It seems, however, that upon their knowledge of how to make food and several other kinds of material depends the existence of all the others. The following from a daily paper seems to be a true expression of the situation: "Few things in science are more startling than the realization that man's existence depends absolutely on tiny vegetables so small that they can only be seen with the most powerful microscope. It is these minute growths which produce the larger vegetables man uses for food and were it not for them the human race would starve to death.

"The origin of life may be a disputed matter, but the operation of life is unquestionably due to microscopic forms of plant life which we call bacteria. There have been many learned definitions of the differences between plant and animal, but few are more satisfactory than that which declares a plant to be an organism that can derive its food from mineral substances and an animal to be an organism that cannot."

Did you ever realize that the cells of every plant or

animal, man included, must keep up a continual fight for existence? Some small species of cells like the diphtheria cells or lock-jaw cells fight our body cells with poison. Their poison is so deadly that the white cells or soldiers of our bodies, who have the business in charge of destroying all the enemies of the body, are compelled to actually fight these diphtheria germs at a distance and apparently also with poison. The following from a reliable medical textbook is interesting:

"Very important members of the cell community are certain colorless cells, the white blood corpuscles or leucocytes, which float in the blood stream or wander in the tissue spaces. These cells are the sanitary inspectors, the police and the scavengers of the community. They can be seen to approach intrusive microbes, inclose them with their own substance and destroy them, apparently by digestion, or else perish themselves, apparently by poisoning. On account of this function of ingesting microbes, they have been termed phagocites. The skin, food and air passages swarm with bacteria all the time and only now and then do they enter and get a foothold. The fact that microbes of disease are often able to break down the defense and enter the tissue, whereas the harmless types are unable to do so, points to the fact that the former have special means of offense and have adapted themselves by special means of offense and defense.

"Some microbes like tetanus and diphtheria elaborate a poison in self defense. The toxin or poison enters the blood stream at the small area affected and produces the symptoms of poisoning. Toxins like pepsin and trypsyn are digestive secretions.

"It appears that phagocites and possibly other cells of the body secrete substances as harmless to themselves as pepsin is to the stomach cells, which act as counter toxin and poison the microbes, and that they are stimulated to this act by the presence of the microbes and their toxins. Presumably those toxins are the same as are used by phagocites to destroy microbes by ingestion. The toxins of all microbes are not alike as shown by different symptoms from diphtheria and tetanus. In acute diseases such as diphtheria and tetanus, though the phagocites crowd towards the infected area until the red and inflamed tissue surrounding it is full of them, killed or paralyzed by the concentrated toxins, they do not ingest the microbes, at any rate, at first. If unable to cope with the invaders, the sufferers die, but in case of recovery they gradually get the upper hand and in the latter stages of the disease the disintegrating microbes may be seen within them."

A most wonderful experiment to show the extraordinarily keen sense of smell possessed by the white cells of our body, who fight and kill the dangerous cells or bacteria, is described in the following from a medical journal:

"If microbes are introduced into the body, inclosed in a capillary glass tube, the ends of which are plugged by a substance that permits a free diffusion of fluids but prevents the escape of the micro-organisms or the entrance of the phagocites, the latter collect about the tube in clusters at the open ends. As sanitary officials, they are attracted by the secretions (toxins) of the microbes."

Just notice how the soldiers of our body we call white cells are able to detect the presence of the dangerous enemies inside the glass tube and wait there at the entrance in sufficient numbers to be able to destroy them if they should escape from the tube. Here we have a good illustration of the fact that the sense organs of the cell are in every way just as keen as those of the higher animals. The cell must know from experience with those bacteria in the past ages that they are very dangerous. His knowledge of the germ tells him that he must not run any chances but destroy it on sight. It is a very singular thing, but statistics show it to be a fact, that the germ cells of consumption will not kill a woman with child, but will always give her a chance to give the child a start in life. This would seem to indicate that the germ of consumption is a friend to the human race. We never stop to realize that the cells are at the base and the cause of all life that we see. A school text on botany has the following to say about the cells that build plants and trees:

"The peculiar work of green plants or green parts of plants is to manufacture the kind of food best known as sugars and starch, such foods being called carbohydrates. This manufacture is exceedingly important, for all life is dependent upon it. If green plants should stop the manufacture of carbohydrates, the food supply of the world would soon be exhausted. All other forms of food are derived from carbohydrates in some way and only green plants can add to the stock that is being drawn upon continually. This means that green plants must manufacture carbohydrates, not only for their own use but also for the use of animals and of plants that are not green." While this statement is true, still we must remember that the plant does not produce anything. It is the cells in the plant that produce in the same manner as it is the men in the factory who are the real producers.

Man has been able to discover but very few of the chemical secrets known by the plant building cells and the other cells who know how to make their own food from the raw material of earth, air and water. The power to produce a light without the expenditure of heat has been known to the cell for ages. Man with the phenomenon before his eyes on sea and land, spending fortunes, investigating this and trying to discover the secret, is not yet able to produce it. A text book on physics has the following to say about it:

"We burn a jet of gas in order to produce artificial light. We are seeking to set up ether waves of a certain length but in doing so we produce only three per cent of these waves and 97% of waves we do not wish and could do very well without, as they are simply dark heat waves. The hotter the body is which is setting up the ether waves, the better is the percentage of useful waves, but even with electric arc lamps we can only attain an efficiency of ten or fifteen per cent. If we could only imitate nature as we see her producing light in the glow worm, where practically the whole of the ether disturbance is in the form of visible light—no dark heat waves being produced—we should be able to cause illumination on a grand scale. Referring to the luminosity of the glow worm, Sir Oliver Lodge has remarked that if we could only obtain this secret from nature, a boy turning a crank could furnish sufficient energy to light an entire electric circuit."

You will notice how they put it. They say if we could "imitate nature" or "discover the secrets of nature." Why not tell the truth and say the secrets of the cell? What is the reason for all this meaningless language and jumbling of words? Whether you say the secrets of nature, the secrets of the universe, the secrets of the world or the secrets of the moon, we understand one about as well as the other. If the cell living a single life in the open, or in a colony like the glow worm or lightning bug, can produce this light, why not be honest about it and admit the truth? There can be no question about the facts. The roots of plants taste, feel, and select from the soil only the material necessary to build the plant and reject all other material. It is also evident that a great number can taste and feel with their leaves as is evidenced by the fly catching plants. Even Haeckel admits that much intelligence and makes the following remarks in reference to it:

"The digestion corresponds to the gastric juice in the animal and is only secreted by the corpuscles if the solid foreign body is nitrogenous (flesh or cheese). Hence the leaves of these insectivorous plants taste their meat diet and distinguish it from other solids to which they are indifferent. In the broader sense, in fact, we may describe the points of the roots of plants as organs of taste. They plunge into the richer parts of the earth, which yield more nourishment and avoid the poor parts."

You see from this that the cells in the plant feel, taste and digest with a gastric juice in precisely the same manner as we do. In fact, after close investigation it has been fully established that all cells whether living singly or in colonies like plants and animals, digest, reproduce, respire and perform all the functions of an animal. They all have to have oxygen and food, the only difference is their method of obtaining it. You can see the same process of social development in all the lower animals, insects and cells, as in man. There is no difference whatever, and on this point Mr. Haeckel makes the following remark: "Division of labor or differentiation, which has just recently begun to be correctly valued, forms a sixth evolutionary function of special importance. We have already seen that division of labor is the strongest impulse towards progressive evolution, not only in civic and social life but also in the social cell confederacy of every many celled organism. A glance at any community or state organization shows that the first condition of higher development and civilization is on the one hand the division of the various duties among the various classes of the citizens and on the other hand, the co-operation of these individuals for the common purpose of the state. This is exactly the case also in every many celled organism. Every multi-cellular individual in the plant or animal kingdom is more perfectly developed and ranks higher in proportion as the division of labor among its constituent cells, the differentiation of its cell individuals is more perfect. Therefore in the various classes of organisms, we find this differentiation, sometimes in a more, sometimes in a less perfect condition. The simplest form of division of labor occurs in those lower animals in the bodies of which only two kinds of cells have become differentiated. This is the case for example, in the lowest plant animal, in sponges and the simplest polyps, as well as in their common parent form, **the gastraea.**

"Throughout the entire many celled bodies of these there are only two different kinds of cells, the one kind affect nutrition and reproduction of the animal, the other kind are its organs of feeling and motion. These two kinds of cells are identical with those which first come to perfection in the first process of differentiation of the germ layers in the human embyro, but in most higher animals, the differentiation of the cell proceeds much further. Some take merely the office of nutrition; others that of reproduction; a third group constitute the outer covering of the body and form the skin; a fourth group, the muscle cells, form the flesh; a fifth group, the nerve cells, develop into the organ of sensation, of will, of thought, etc. All these different kinds of cells originally proceeded by differentiation or specialization from the single egg cell and from the homogeneous descendents of that egg cell, owing to division of labor. This differentiation of the cells or this division of labor originally arose in tribal history from causes similar to the division of labor in the civilized states of men. Afterwards it appears in the germ history and by that time it has been made over to heredity and is merely repeated in accordance with the fundamental law of biogeny."

The reader will notice that Mr. Haeckel attempts to make it plain that the cell has developed from his single separate savage life to the high state of civilization and organization that we now find him, in plants and animals in precisely the same manner as the human race. No one can conceive how the cell or insect could have developed their social institution without intelligence of the same nature as man. It is simply foolish to try to evade plain common sense and attempt to bewilder us with meaningless phrases. Our ordinary experience and common sense is bound to bring us to these natural conclusions because they stand out as the simple naked truths and facts.

Every plant or animal starts from one single cell, which multiplies and as the body grows and the organs are completed in size and structure, we have these groups of cells known as organs or cells of the brain, cells of the liver, cells of the skin, cells of the muscles, of the bone, etc. We see that the cells of the body are of the same family, but by reason of their different occupations they have become a little different in general appearance in just the same manner as the children of a family, who would occupy different positions in life, would by virtue of their difference in occupation look a little different. If one was a book-keeper, one a butcher, one a farmer, one a soldier and one a preacher, they would all look a little different in general appearance. Every animal will become adapted to the particular work he has to do and this we find is also the case with the cells as well as man. This being the situation, the cells must and do possess the same intelligence. That they all possess the same general knowledge and skill is evidenced by the fact that they can do such work as may be necessary in any particular situation, as for instance in the case of repairing broken limbs or cut tissues. As I said before, this work required in any particular place just as much intelligence as building the part anew. Any work in life requires intelligence; the blacksmith, clerk, butcher, lawyer, all must use intelligence in their work, and so it is with the cells of our body. The intelligence may differ in degree in the same manner that there are good and poor lawyers and blacksmiths, but to be a lawyer or blacksmith at all will require intelligence.

Can the cells of muscle and motion perform the work of sensation or direction? Can one cell do the work of the other? We find that they can, but not as well. Take for instance in the case of the polyp and polyzoan which in every way are just alike in appearance, but one has cells which we call the nervous system, for the special purpose of keeping watch and giving notice of the approach of enemies, while the other has not. The polyp when touched will attempt to escape but, as it has the work of notifying all the other cells to contract, its rapidity of contraction is interfered with. The cells of the polyp, although they can do the work of the nerve cells, cannot do it with the skill and rapidity that the nerves do it, who specialize in that work and have no other work to do. We find this to be the case with animals and men in exactly the same degree. In reference to ,this Spencer states: "A polyp and a polyzoan, two similar in outward appearances, but very unlike in their internal structure, will serve for comparison. A tentacle of a polyp when touched slowly contracts, and if the touch has been rude, the contraction presently extends to the other tentacles and eventually to the whole body, but if you touch a tentacle of a polyzoan or slightly disturb the water near it the whole cluster of tentacles is instantly withdrawn, along with the protruded parts of the creature's body, into its sheath. The one creature has no specialized contractile organs or fibers for conveying impressions. The other has definite muscles and nerves."

This should illustrate clearly what a nerve cell really is and that the cells of a structure can perform different and various kinds of work and do so until they discover the better and more effective way of doing it, which is to specialize and allot to each one his particular and specific work. The polyzoan, however, has no brain, simply a nervous system, that is, cells strung all over and through the body who have nothing else to do but watch for enemies and direct the actions of the individual.

We have seen that the cell is a perfect animal, having all the attributes of the highest developed animal. Ages before the social cells discovered the advantages of associating together for mutual protection and assistance, they lived separate lives in different tribes and classes, each one adapting itself to the particular climate or conditions to which it happened to become exposed. It did this just as plants, animals, and man do today. Just as man is the cause of all the structures on earth produced by him, such as railroads, armies, navies, cities, etc., so the cell is the cause of the living things that we see. The power in man to do these things is his intelligence, and in the same manner it is intelligence that gives the cell his power. The structures produced by man are based on centuries of accumulated experience and so it is with the cell. He also must act in accordance with the knowledge accumulated in the past ages.

Although Mr. Drummond does not see life in the same light that I do I shall quote his impressions of the situation: "Now all these complicated contrivances, bones, muscles, nerves, heart, brain, lungs are made out of cells. They are themselves in their furthest development simply masses of centers of cells modified in various ways for the special department of household work. They are meant to serve. No new thing has entered into the embryo since its first appearance except building material. It seized whatever matter lay at hand, incorporated it with its own quickening substance and built it into its appropriate place, so the structure rose in size to the stature of man. The immense distance man has come between the early cell and the infant's formed body, the evolutionist sees concentrated into these few months, representing the labor and progress of ages. Here before him is the whole stretch of time since life first dawned upon earth. The human form does not begin as a human form, it begins as an animal. At first there is nothing wearing the remotest semblance of humanity; what meets the eye is the vast procession of lower forms of life. To-day in the embryo of still living things, we find again a resurrection and life in the frame of man himself. The proposition is not only that man begins his earthly existence in the guise of a lower animal-embryo, but that in the successive transformations of the human embryo there is produced before our eyes a visible, actual, physical representation of part of the life history of the world."

This description of the situation by Mr. Drummond is both very good and instructive. It shows how the cells in the course of construction of the human being are compelled to follow in their path of past experiences.

The human being evolving from the cell to the new born babe takes on all the forms of the past life of man for ages. The size of the structure does not make it any less complicated and difficult to produce. Take for instance the spinning apparatus of the spider,—the spider's weaving machine is considered the most intricate and perfect device in the world for its purpose. It has over 600 spools and bobbins so small that they can only be seen with a microscope, magnifying about 200 diameters. The threads which the spider produces are formed in the following manner: the raw material for the web is a cellulose mixture and is contained in sacks; muscular pressure forces this liquid through minute ducts and this liquid is so made that it immediately hardens into a fiber when exposed to the air. This fiber is brought to the desired degree of thickness and thinness by winding over and under the various spools and bobbins. He has claws made specifically for handling his threads and spinning machinery and with these specially made hands he is able to handle his machinery with wonderful skill and speed. He is able with this machine to produce threads of different degrees of strength, depending on the purpose for which they are to be used. The silk worm uses a similar method, but his is crude compared to the spider's. The material used, however, is almost the same. Man has been experimenting until he discovered the substance or raw material used by the cells in the silk worm and is now able to imitate it tolerably well and produce an artificial silk fairly good.

Several Germans and Frenchmen are experimenting with spiders, trying to breed a special kind who might be a competitor of the silk worm. Just think of what the spider building cells have to do to build this spider! The building of a tree, man or elephant is simple as compared with the building of the spider. Think of a spinning machine with 600 spools and bobbins and the dexterity required to handle it, and the job of manufacturing the cellulose from other crude material with which to weave the threads to be used for snares and other purposes! Where does man produce any such rapidly moving, complicated factory? Until a person has seen some of the productions of the cell through the microscope, he does not comprehend what life is. A certain species of the spider, we remember, weaves a waterproof diving bell, which he ties to the grass under water, and in which he dwells with his family and catches water insects for food when they come near him. He carries the air from above down into his underwater dwelling, which is both air and water-tight.

The intelligence of man cannot produce any structure involving as many complicated acts of skill and design as the spider, nor can man show any more social progress than the bee. There is no doubt that the bees have some method of simple and rapid inter-communication. If anything happens to the hive or if some new place to get honey is discovered, they all know about it very quickly. Mr. Maeterlink has the following to say about this:

"Let us now in order to form a clearer conception of the bee's intellectual power proceed to consider their methods of inter-communication. There can be no doubting that they understand each other; and indeed it were surely impossible for a republic so considerable, wherein the labours are so varied and so marvelously combined to subsist amid the silence and spiritual isolation of so many thousand creatures. They must be able therefore to give expression to thoughts and feelings by means either of a phonetic vocabulary or more probably of some kind of tactile language or magnetic intuition corresponding perhaps to senses and properties of matter wholly unknown to ourselves. And such intuition well might lodge in the mysterious antennae—containing in the case of the worker, according to Cheshire's calculation, 12,000 tactile hairs and 5,000 'smell-hollows', wherewith they probe and fathom the darkness. For the mutual understanding of the bees is not confined to their habitual labours; the extraordinary also has a name and place in their language, as is proved by the manner in which news, good or bad, normal or supernatural, will at once spread in the hive; the loss or return of the mother, for instance, the entrance of an enemy, the intrusion of a strange queen, the approach of a band of marauders, the discovery of treasure, etc."

It has been shown that bees take advantage of every invention and discovery which they may run across and that they use their building material or food to the very best advantage. In reference to this matter, Mr. Maeterlink has also the following to say:

"Scarcely had it been formulated when another naturalist, Andrew Knight, having covered the bark of some diseased tree with a kind of cement made of turpentine and wax, discovered that his bees were entirely renouncing the collection of propolis and exclusively using this unknown matter, which they had quickly tested and adopted and found in abundant quantities ready prepared in the vicinity of their dwelling.

And indeed, one-half of the science and practise of apiculture consists in giving free range to the spirit of initiative possessed by the bees and in providing their enterprising intellect with opportunities for veritable discoveries and veritable inventions. Thus, for instance, to aid in the rearing of the larvae and nymphs, the beekeeper will scatter a certain quantity of flour close to the hive when the pollen is scarce, of which these consume an enormous quantity. In a state of nature in the heart of their native forests in Asiatic valleys, where they existed probably long before the tertiary epoch, the bees can evidently never have met with a substance of this kind. And yet, if care be taken to 'bait' some of them with it by placing them on the floor, they will touch it and test it; they will perceive that its properties more or less resemble those possessed by the dust of the anthers; they will spread the news among their sisters and we shall soon find every forager hastening to this unexpected, incomprehensible food, which in their hereditary memory must be inseparable from the calyx of flowers where their flight for so many centuries past has been sumptuously and voluptuously welcomed."

The brain of the bee, which directs all its actions, contains but few cells. The power that nearly all caterpillars or young butterflies have of changing their color to harmonize with the place in which they are living is certainly wonderful, and its value has not been understood by man until lately. During this last European war, the following article appeared in my paper on the subject: "A striking instance of the application of scientific knowledge to the purpose of war is the color of the German service uniform, a kind of invisible grey-yellow-green, which blends with the prevailing hue of a landscape, so that bodies of troops become as it were concealed in a chromatic haze.

"Naturalists have long been familiar with 'protective coloration', which is found among many species of insect and other animals, and it is rather surprising that this principle, which science declares to be a result of evolutionary adaptation, has not sooner been employed in the apparelling of armies. Khaki uniforms, which have been widely adopted since the Boer war, represent a first step toward the development of a scientific military investment. . . . . . Evidently a dress that tends to conceal the movements of troops is almost as important as smokeless powder, which keeps the secret of the location of guns, and its importance is accentuated by the increased distance at which armies now fight one another. Troops dressed in dark or flaming colors can be seen afar off, but the modern scientific uniform possesses something of the concealing power of the chromatic pattern with which leopards, tigers, and zebras are covered.

"Those who have never seen these animals amid their natural surroundings can form no clear idea of the blinding effects produced by protective coloration. Hunters in Africa have told almost incredible stories of herds of zebras standing in plain sight of the observer and yet actually invisible to him until a sudden alarm sent them away in wild scamper. Their color and their markings are wonderfully accordant with yellowish hues and the barrings of black shadow that characterize the landscape amid which they live.

"In our day the Assyrian wolf no longer comes down upon the fold, his cohorts gleaming in purple and gold, but he steals along like the tawny lion in the tawny desert, while his dusty hued prey slips off concealed against the background of sand.

"These things may have a determining influence in disgusting mankind with war. Mars in the guise of a chameleon loses all his romantic attraction and shows himself up as a hideous monster, whose doings become more repulsive in proportion as they are more brutally practical."

Figs. 47-49 and 50 and others by Mr. Shute are published in this book by courtesy of the Open Court Pub.

Co. These figures show the wonderful skill with which the cell colonies in butterflies, beetles, etc., are able to build their moving structures or habitations in shapes and colors so as to deceive and escape their enemies. Fig. 47 shows that they have in addition painted the inside of the wings a most beautiful color to attract the opposite sex. Intelligent man is doing the same thing today, but he did not discover the art nor its benefits until just recently. The cells who build these beetles, caterpillars and butterflies have understood and practiced the art for ages. They knew the secret of how to produce the coloring matter and with it paint these artistic figures thousands of years before man lived in houses.

I have myself experimented with caterpillars, who can color themselves as may be necessary to simulate the place on which they are resting and it certainly is wonderful when you consider the situation, that the cells of the skin must first have a picture of the outside before they can arrange the pigment so as to affect the desired color. Just as wonderful is the ability of the cells, which make up the caterpillar, to tear down and dismantle the caterpillar and with the same material build a new and different structure designed to navigate the air, which we call a butterfly. A text book on zoology has the following on Protective Coloring: "Mr. Leslie inclosed certain caterpillars of one kind in two boxes, one black and the other white, and he found that the color of the chrysalis in each case harmonized with the color of the box. Mr. R. Holland also found the cocoons of the Emperor moth to be either white or brown, according as they were spun on paper or amid dead grass, or on soil. Mr. E. B. Poulton has ascertained that in a large number of larvae of a Vanessa butterfly surrounded by variously colored papers, the colors of nearly all the pupae were like, or related to that of the paper about them."

In what manner do the actions of the cells in animals, taking advantage of the color protection, differ from the same acts by a German general? In what manner does the act of the bee going after and storing food for the winter differ from the act of the farmer filling his barn with hay for the winter? They are all directed for a purpose. At every step in our investigation of the cause of evolution, development and progress in life, we shall find that it is intelligence in the cell that is the cause. It is not chemical nor mechanical force nor blind chance, as some will have it, nor is it nature, destiny or some other mystic force as others call it. It is the cell; and the reason he is able to produce these different things is the same as the reason that man is able to produce those different things produced by him. We find that he has taken advantage of all the blind forces of nature and converted them to his own purpose and use, just as man has. Only a few years ago we knew nothing about this invisible architect and we could only say that a tree or man simply grew, but now we know there is one who directs the work in the building of a tree or animal, just as there is one who directs the work in the construction of a house. The structure goes up piece by piece. The materialist,physicist, or mechanist states that when certain conditions get right, the chemical force will do the rest and produce all the different living structures we see. This statement we know is not true from what we know at the present time. Nothing is produced today, nor has anything been produced in the past, unless some intelligent being was back of it to direct the dead matter and blind forces in the universe.

The natural and chemical forces applied to matter will

always produce the same result and will always follow fixed laws. We know that back of all those things we see like houses, factories, ships and railroads, is the intelligence of man. We see the cell build analogous purposive structures we call plants and animals, and we are compelled to say that back of them is the intelligence of the cell. If it is necessary in one it is also necessary in the other. We are not able to say or explain the cause of intelligence in the cell any more than we can explain the cause of electricity and gravity. However, we know that its existence is a fact. We know it is intelligence that guides the actions of man, animal and cells just as we know it is gravitation which guides the stone towards the ground. We find that it is the same force in all places which guides them in their acts, just as it is the same force that pulls all objects toward the ground. We find that the hydra can be turned inside out without perceptibly injuring the animal and that in that case the cells will change work. The outside skin layer of cells, which before did all the respiratory work, will now do all the digesting, and all the inside layer of skin cells will do all the breathing or respiratory work. What does it prove? Simply this,—that the cells are intelligent,—they know what their duty is and what must be done and know how to do the right thing at the right time under all circumstances. In the larvae of the dragon-fly and a number of other insects, the alimentary canal respires, digests and excretes. It might seem singular that the stomach cell can do all these different kinds of work, but when we consider that the larvae or insect is simply a colony of individuals who work for themselves and do whatever is necessary to perpetuate and protect their colony, it is only what we may reasonably expect. One will take charge of and do the work of the other or all the work when it may be necessary under certain circumstances.

These facts simply show that the actions of the cell are in every detail similar to those of man, and they show that chance variation is not the cause of producing anything in life any more than chance would produce anything like houses and railroads, and that chance variation is in no way the cause of evolution and progress. Take for instance the electric organs of fishes used by them both for offensive and defensive purposes. They are very large and until fully completed they could be of no use to the individual. It could not have been of any benefit to the fish to carry that large complicated electric mechanism around with him until perfected to such an extent as to be used. It could not possibly have been a beneficial variation until completed. The following is a description of one of these fishes from a scientific magazine: "The marshy waters of Bera and Rastro in South America are filled with innumerable electric eels, which can at pleasure discharge from every part of their slimy, yellow-speckled bodies a deadening shock. This species of gymnotus is about five or six feet in length, and is powerful enough to kill human beings and the largest animals when it discharges its nervous organs at one shock in a favorable direction. It was once found necessary to change the line of road from Uritucu across the steppe, owing to the number of horses which in fording a certain rivulet annually fell a sacrifice to these gymnoti, which had accumulated there in great numbers."

The production of these powerful electric machines is nothing more than what we should reasonably expect when we consider who the builder of the electric fish is and his mental and inventive capacity. If you should examine the cell that builds electric fish with a powerful microscope, side by side with the cell that does the thinking for the great inventor, Thomas Edison,—compare them critically and closely, they would appear to you to be one and the same cell and the one could not be distinguished from the other.

There is a plant in Persia called the Devil plant, that has invented a scheme by which it is able to capture deer, cattle and animals by the nose and kill them with poison, and when the animal dies the plant grows in the soil made rich by the decaying carcass. This plant (or rather the cells that build it) makes a pair of spring hooks about 6 inches long and 4 inches apart at top, so as to fit over the nose of the animal. The hooks spring inward and are tough as steel. The following is a brief description of the plant:

"The Devil plant they call it in Persia, and well they may, for it is more deadly to the flocks and herds that play so important a part in their life than is the loco weed to the herds of our southwestern states.

It is in the fall that the Devil plant gets in its deadly work. The flowers give place to seed pods with great belly like capsules and long stiff claws like those of a beast of prey. These are hidden under the brown and yellow leaves and when a grazing animal—a sheep, a camel, a wild ass or an antelope for example—browses among the foliage, the claws hook themselves into its nostrils. The animal tries to rub them off but the more it rubs the deeper it forces the claw like hooks into its skin. Its throat becomes so inflamed that it can neither eat nor drink and consequently it dies of starvation and pain. The animal's body lies in the open and decays and into the decomposing flesh the hundreds of black seeds contained in the capsules of the clawed pod are discharged. For it seems that earth is not rich enough for them and only in decayed flesh can they find enough nourishment. Drivers of caravans curse the Devil plant

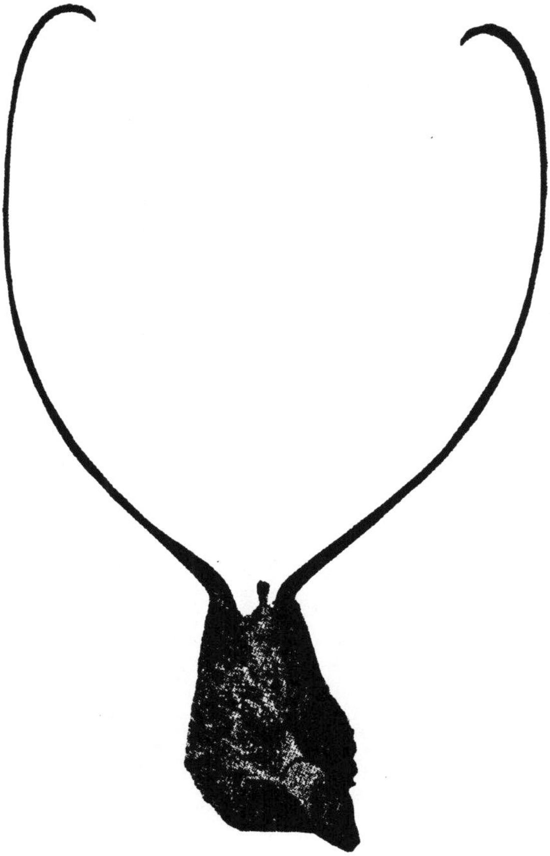

FIG. 48.—Seedpod of the Persian devil plant, with its claws four inches across

for it may cost them many of their beasts when these are turned loose to graze at night. But most of the semi-wild beasts that graze in the country have learned to avoid it, even as the American herds have learned to avoid the deadly loco weed."

Do you think a scheme like that could have come about and have been produced by chance? Before they could in any way become effective, both of those hooks had to be just so far apart and curved over just so and made of extraordinarily strong material to effect the purpose for which they were intended. This grappling machine would have to be complete and perfect in every way before it could be effective. It seems almost unreasonable to think that a plant could figure out this scheme with which to be able to catch, kill and eat cattle, sheep and camels, but such are the facts. This plant is a stationary abode of a colony of cells. The cattle, sheep and man are movable abodes of cell colonies. The cells of a stationary colony should have just as much time and opportunity to figure out schemes and inventions with which to protect themselves or obtain their food as the cells of the movable colonies. When the matter is considered rightly, we should expect to find precisely the same inventive genius, skill and intelligence in one place as the other.

Before closing this chapter I shall quote the following by Ernest Haeckel from his works on Embryology. It is a little long but it is a good comparison of the human cell with other cells living singly and as separate lives. He says:

"Though the amoeba is therefore only a simple cell, it shows itself capable of performing all the functions of a many celled organism. It moves itself by creeping, it feels, it feeds, it reproduces its kind. Some species of amoeba are visible to the naked eye, but the greater number are microscopic. Our reasons for regarding the amoeba as the particular one celled organism, the phylogenetic relations of which to the egg cell are of peculiar importance, will be evident from the following facts. In many lower animals the egg cell remains in its original naked condition until it is fertilized; it requires no covering and is often indistinguishable from an amoeba. Like the latter, these naked egg-cells can extend processes and move about. In the sponges these active egg-cells creep freely about as though they were independent amoeba, even within the parent organism. In this condition they were observed by earlier naturalists and were mistaken for amoeba, living as parasitical intruders in the body of the sponge. It was only afterwards that it was discovered that these supposed one celled parasites were in reality the egg cells of the sponge itself. This remarkable phenomenon is also found in other lower animals—for example—in those pretty bell shaped plant animals (medusae); the eggs of these also remain as naked uncovered cells, which stretch out amoeboid processes, feed themselves, move, and from which after fertilization the many celled medusae-organism is indirectly or directly developed by repeated division.

"It is therefore certainly no wild hypothesis but an entirely sober conclusion which regards the amoeba as the particular one celled organism which gives us an approximate representation of the ancient one-celled ancestral form common to all many celled organisms. The naked simple amoeba possesses a less differentiated and more primary character than most other cells. To this may be added the circumstance that similar amoeboid cells can be shown in the full grown bodies of all many celled animals. For example, they occur as the so called white blood corpuscles among the red blood cells (corpuscles) in human blood and in that of all other vertebrates. They also occur in many invertebrate animals, for instance, in

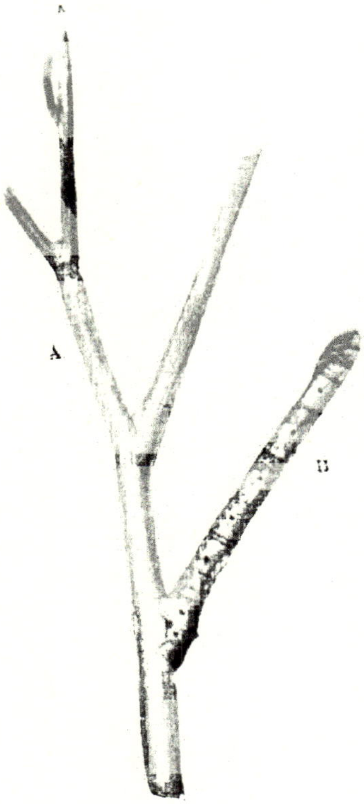

Fig. 49.—Caterpillar B of a Geometer Moth (Prochoerodes transverrata) on the stem of a plant (Ailanthus) A. Illustrating protective resemblance in both color and form. Schute.

the blood of the snail and in 1859 I showed that these colorless blood corpuscles like independent amoeba can assimilate solid particles, can, therefore, eat. Lately it has been found that very many different cells, if they have room, are able to move and eat and to act entirely like amoeba.

The capacity of the naked cell to make these characteristic amoeboid movements depends on the contractility (or automatic movableness) of the protoplasm. This seems to be the universal property of all young cells. Where they are not surrounded by a strong membrane or shut up in a cell prison, they are all capable of amoeboid movements. This is as true of the uncovered egg cell as of other uncovered cells of the moving cells of various kinds, lymph cells, mucous cells, etc.

"Our examination of the egg cell and comparison of it with the amoeba has afforded us the best and surest basis for the history of the germ as well as for the history of the tribe. From it we have drawn the conclusions that the human egg is a simple cell; that this egg cell is not essentially different from those of other mammals and that we must therefore infer the existence of a primeval one-celled ancestral form, which in all essential points was of amoeboid form.

"The very important bearing which the Cell Theory has on the whole conception of organic nature is thus very clearly seen. The 'Place of man in nature' is radically explained by it. Without this theory, man is an unintelligible puzzle. Philosophers, therefore, and certainly the psychologists, ought especially to acquaint themselves thoroughly with the Cell Theory. The human mind can only be really understood by means of this theory and its simplest form is illustrated in the amoeba. The extant amoeba and the kindred one-celled organism, arcellae gregarinae, etc., are therefore of great interest, because they show us the simple cell in a permanently independent form. The human organism and that of higher animals, on the contrary, is only one cell in its earliest immature condition. As soon as the egg cell is fertilized, it multiplies by division and forms a community or colony of many social cells. These differentiate themselves and by their specialization by various modifications of these cells, the various tissues which compose the various organs are developed. The developed many-celled organisms of man and of all higher animals resembles therefore, a social, civil community, the numerous single individuals of which are indeed developed in various ways, but were originally only simple cells. * * *

"We have reached the conclusion that the original, ancestral form of man, as of the other animals, was a one-celled organism. The whole difficult problem of the History of Evolution is thus now reduced to the simple question: How has the complex many celled organism arisen from the simple one celled form? By what natural process has the simple cell been transferred into the complex life apparatus with all its various organs, the apparently rational and purposive construction of which we admire in the developed body?

"Turning now to answer this question, we must bear in mind the view to which we have already alluded, that the many celled organism is ordered and constituted on the same principle as a civilized state in which the several citizens have devoted themselves to various services directed towards common ends. This comparison is of the greatest service in enabling us thoroughly to understand the construction of man from many cells of various kinds and to understand also the harmonious co-operation of these various cells for an apparently preconceived purpose. If we bear this comparison in mind and apply this significant idea of the developed many celled organism as a civil union of many individuals to the history of the evolution of this organism, we shall obtain a correct view of the real nature of the first and most important processes of evolution. We can even on deeper reflection guess the first stages of development and establish them a priori, before we call observation, a posterior knowledge, to our aid.

"Let us therefore first answer this question: 'Granting the correctness of the fundamental law of Biogeny, how would the original one-celled organism, which founded the first cell state and thus became the ancestor of the higher many celled animals—how must that organism have acted at the beginning of organic life on the earth or at the beginning of creation, as it is usually expressed?' The answer is very simple. It must have acted just as man who founds a state or a colony for a given purpose.

"Their only purpose in life for centuries has remained as simple as that of the lower animals or plants; the simple aim of self preservation and of the production of descendants. They have been contented with the simplest organic function, nutrition and reproduction. Hunger and love are their only motive for action. For a long period these savages must have aimed at the one single object of self preservation. Gradually, however, several families collected at certain places—larger communities arose and now many reciprocal relations began to arise between individuals; in consequence a rude division of labor took place. Certain savages continued to fish and hunt, others began to cultivate the ground, others devoted themselves to religion and medicine, which now began to develop, etc. In short the ever increasing division of labor specializes the people into various ranks or castes, which always tend to become more sharply defined in proportion as the state became more highly developed; all follow diverse occupations and yet work for a common end. A process similar to this, and the details of which each can easily fill up for himself, took place millions of years ago when, at the beginning of organic life on the earth, one celled organisms at first developed and were afterwards followed by many celled forms.

"The single cells which arose by reproduction from the oldest parent-cells must at first have lived in an isolated condition; each one performed the same simple offices as all the others. They were satisfied with self preservation, nutrition and reproduction. At a later period isolated cells gathered into communities. Groups of simple cells, which had arisen by the continued division of a single cell remain together and now began gradually to perform different offices in life. The first traces of specialization or division of labor soon occurred, as one cell assumed one office, another another. One set of cells may have devoted themselves especially to the absorption of food or nutrition. Other cells may have busied themselves only with reproduction and others again have formed themselves into protecting organs for the little community, etc. In short, various classes or castes must have arisen in the cell state following diverse occupation and yet working together for the common end. In proportion as this division of labor progressed, the many celled organism or the specialized cell community became more perfect or civilized."

Mr. Haeckel is a great student of life but he does not think that the cell is possessed of any intelligence, but believes that the actions of the cells are caused by chemical energy. I think he is mistaken,—at least, all facts that we know so far are against him. He makes this statement: "How must that organism (meaning the cell) have acted at the beginning of organic life on the earth or at the beginning of creation, as it is usually expressed? The answer is very simple. It must have acted just as man who founds a state or a colony for a given purpose."

It seems to me perfectly clear that if the cell organized his civilized communities and colonies in the same manner as man did his, that he should also be and necessarily was an intelligent being like man. If man is able to do these things by reason of his intelligence, the same reason or cause must also be back of the cell. Our knowledge of the cell is, as yet, very limited, but if man is intelligent so is the cell that built him. If the cell can invent and produce machines for different purposes and organize himself socially in the same manner as man does, why is he less intelligent? I want the reader to carefully consider the following statement by Mr. Haeckel in describing the actions of the single cell called amoeba, which is so very similar to the man and animal building cell. He states:

"If one of these creeping amoeba is touched with a needle or if a drop of acid is added to the water, the whole body at once contracts in consequence of this mechanical or chemical irritation. Usually it reassumes its spherical form. Under certain circumstances, for example if the impurities remain in the water, the amoeba begins to encase itself. It exudes a homogeneous envelope or capsule, which immediately hardens and in a state of repose assumes the form of a spherical cell surrounded by a protecting membrane."

Now what would you do under similar circumstances? If someone should punch you with a crowbar or soak you in acid, would you not also become irritable, would you not begin to think it was time to cover yourself with a protecting membrane? That is just what the amoeba does. He understands how to make and carry a material along with him, with which he is able to cover his body and protect himself when necessary under certain dangerous circumstances like this. To be in position and to be able to provide for such emergencies shows an extraordinary degree of intellect and foresight. A material must be discovered and made and carried ready at hand, which when exposed to the air immediately hardens into a protective shell. An animal who can conceive the idea, discover and make the substance and use it when necessary, as does the amoeba, certainly shows skill and foresight of a high degree. We have no reason to doubt his intelligence, for as Mr. Haeckel himself states, he looks and acts in every way like the animals that do the thinking for us, so if Mr. Haeckel is intelligent, so are the cells.

When the cells build a plant or animal they proceed in the same manner as man in building a structure. They build all the parts with a view of combining them all into one whole. The parts are made to work in combination with every other part. The organs of the body are in that regard like the skillful combinations found in the arts and industries of man. The diversity of functions and division of labor begin with the production of organs in order to take care of the different wants of the cell colony. To satisfy these wants certain labor and work must be done and each organ is a delegation to do certain and special kinds of labor. The cells working together in the individual can build organs to a certain extent and purpose with the material at hand and man can extend or increase the use of these organs with other material obtained from the outside,—as for instance, the efficiency of the hand is increased by tools of different kinds for digging, striking, etc. The cell builds teeth with which to grind the food. Man invented another scheme on a larger scale for the same purpose, the mill. The cell made eyes with which to see. Man extended the eyes' efficiency so as to see farther and smaller objects, by the invention of microscope and telescope. The cell builds organs of locomotion to move over the ground. Man invents better and quicker methods by the use of the railroad and automobile.

We know how and why we build and so we should know how and why the cell builds because we are one and the same. In our memory we have a record of past events and experiences. This record constitutes what we know—it is our knowledge and our intellect. From this past record of experiences we form judgments, which determine our actions. Every being must act from experience and not otherwise. The knowledge and experience of the human cell is only that of building the human individual and in the memory of the germ cell there is a record of the experiences of building and guiding the actions of countless individuals in the past together with the record of new and further experience gathered from the life of the individual from which it came. That all the cells of our body have a memory similar to our thinking cells is clearly proven by the fact that we can teach our limbs to perform acts like playing a piano, walking, etc., and when they have learned to perform these acts they will perform them without troubling or demanding any attention from our thinking cells or consciousness. By consciousness I mean those cells in charge of our sense organs, which we also call us, or ourselves. Every cell must be a conscious and intelligent being as well in one place of the body as in the other. If certain actions are performed often enough, so as to be a prominent part in the records of the individual's past experience or memory, those actions will likely reappear in the new individual. We do not understand how the cell is able to take and keep a record of past events for future use (which power we call memory) but we know this to be a fact. From this fact of memory, the power to use the records of the past to guide the actions in the future, we are able to clearly understand the cause of heredity, evolution and progress. Why does a kernel of corn grow into a corn stalk and not into a sunflower? Just for the same reason that a carpenter will build houses and a watchmaker will build watches, and just for the same reason that a muskrat will build huts in which to live over the water while the squirrel will build his nest in the tree. Stored in the memory is a record of past experiences of each of those individuals. The corn building cell has had no experience in building the sun flower nor has the squirrel had any experience in building houses over the water, nor has the carpenter had any experience in building watches.

Many meaningless words have been invented,—for instance, such as geotropic and heliotropic, which are used to express certain properties of plants; that is to say, the roots of plants are said to be geotropic because they grow downward into the earth, while the stem and branches are said to be heliotropic because they grow upward into the sunlight. Now the fact is, there are no such things. The root is not compelled to grow down into the earth nor the stem to grow up towards the sunlight by reason of anything except the own free will of the cells or builders. The cells building the roots are working for a purpose and that is to get down into the ground to get the minerals for building material and the cells building the stem must get up into the sunlight, to get the heat which they must use to effect chemical action in the manufacture of the different kinds of building material and food. All the cells of the plant, be it root or stem, understand their business and know what they are about and what they want. This is clearly proven by the fact that a piece of root placed in the ground will grow in both directions and so will a piece of stem or branch. The cells in either place in the root or stem know what they must have to develop their colony, which we call a plant. A plant is a vast colony of individual cells, working together to perpetuate their existence. To build their habitation called a plant, they must have building material, and this they produce from mixing chemically and mechanically the minerals of the earth and the elements contained in air and water. The cells of the plant building the root going down into the ground for minerals are no more geotropic than a professor going down into his cellar for a can of sour milk; nor are the cells building the stem up into the sunlight any more heliotropic than a preacher going up into his garret for a piece of dried mutton. They all act for a purpose.

Such is the case also with the polyps. Any part of them will grow into a complete polyp. It is clear that in certain cases each cell from the entire colony of cells knows how to build the entire animal or plant. We find this to be the case until the animal or plant becomes very highly organized and specialized, and then we find certain cells are set apart for the specific purpose of building the individual at the proper time, and these are given, as you might say, a special training and college course in the business.

The industries of the bee and their social organization is so very similar to that of the cell that they can be used to illustrate the actions of the cell in many ways, especially to show the same intellect in the smaller beings or structures as in the larger. For instance when the bees are taken south to perpetual sunshine they quit storing honey for the winter. Whole communities of bees sometimes take to theft and live by robbing other hives, just like man plundering his fellow man, and they will destroy the queen first, so as to create a disturbance and disorganization among the workers.

The ants rob and plunder each other, take each other captive as slaves and compel the captives to do all the work while the others live in idleness and luxury. It is perfectly natural that we should find these acts purposive and intelligent to the same degree in the ants and bees as in man, because both man and ant are put together and guided by the same individual, the cell. The cells are individuals that build just as man does and they use the same matter and forces in the production of their different structures. Man has been and is today studying the various building schemes and methods for production and protection now used by the cell.

The cell is the architect, builder and guardian of all living things we see such as plants, animals and man, in the same manner as man is the architect, builder and guardian of all those things we see produced by him, such as houses, railroads, machines, factories, etc. Because man cannot produce a microscope powerful enough to see how the cell produces all his different structures and watch his organs perform their various functions, there is no reason to deny the facts because we know that matter everywhere in the universe follows fixed laws whether it be a molecule, atom, grain of sand, stone or brick. Heat, light, electricity, gravitation, chemical affinity, also obey fixed laws. This we find to be the case everywhere in the universe. The cell interferes with the laws of matter and force just as man does, and as a master of force and matter, he compels them to serve him. The cell colonies we call electric fish gather and store electricity in precisely the same manner as man. The cells of the oyster build strong houses of the microscopic lime and mineral matter that they gather in the water and in order to produce the required hardness they are compelled to employ the aid of chemistry, like man.

There is precisely the same distinction between a cell and other microscopic dead matter as there is between a man and a stone. The cell is a thinking, conscious, intelligent being like man. How can he be anything else when we consider the fact that we are but a colony of these beings? There is no cause for dividing life into animal and vegetable. There are but two kingdoms, the living and the dead. The vegetable and animal start alike and are all built by cells. The cell colonies produce those structures that will be necessary to their existence and which come within their knowledge and experience, as man does.

The cells manifest all the functions of the highest life like eating, choosing, digesting, secreting, working, excreting and generating. Vegetable building cells do just the same thing as animal building cells as far as the functions of life are concerned. Their knowledge and experience and ability to produce things are of course different, as in animals and man. Every cell accommodates and adapts himself to the work to be done, as man does. When not subject to pressure the cell is round, but when subject to pressure he takes on such forms as allow the most freedom and give the most efficiency in accomplishing his work. Thus we see him everywhere manifesting the same reason and actions as man. The male cell of animals in his preparation to meet, and in his effort to find the female cell shows the same wisdom, foresight and intelligence as the smartest general in the army. The centrosome or general manager of the cell body in his careful management of the division of the cell body shows the same foresight and intelligence. His forces are wisely conserved and correctly used. In the face of these clearly and wonderfully intelligent acts, it is absurd to use such words as "instinct," "heliotropism" and "Natural Law." Anyone who can see any explanation in such words and phrases is clearly lacking in the power to correctly observe things. It is just as sensible to say that man is able to produce his battleships, etc., by reason of his "instinct" or that he goes to a banquet by reason of his "heliotropism or that he writes a book by reason of natural law." The difference between the living and the dead is clear and certain. The science of chemistry has proven that atoms and molecules of matter must obey fixed laws in just the same manner as the larger particles of matter like grains of sand or stones and bricks, that everywhere in the universe matter does, always has and always will follow fixed laws.

Based upon these facts man can now produce a large number of things contained in the body and found in living structures, such as carbon compounds and many others. We must, however, not lose sight of the fact that these are not the cell itself, not the living beings themselves. These chemical compounds are the productions of the cell for his own use and purpose. We must remember that the cells cannot produce them for his own use in any different manner than man. The cell or man must follow and observe the fixed laws of matter and force in order to be able to produce the particular article desired. The same laws must be observed in handling atoms and molecules as in handling mortar, bricks or electricity.

The same reason, consciousness, and intellect must be there to guide matter and force in the microscopic world as well as in our world. Knowing and observing the laws of matter, man can mix certain materials like cement and sand and produce a concrete, in similar manner the cell, knowing and observing the same laws, can mix lime and other materials and produce a shell, and knowing and observing the laws of matter, man can make a substance that, when forced through capillary holes, will harden when exposed to the air and when wound into threads will be artificial silk. The cells in silk worms produce the real thing in exactly the same manner. We imitate and follow the process used and observed by the cells in the silk worm, step by step. The production of the silk, whether done by the cell or man; involves a knowledge of chemistry and mechanics and requires intelligence.

Each in his place could produce nothing without knowledge, experience and intellect. I could go on indefinitely in this manner and compare the action and production of man and cell and show the exact similarity. The hard frames of trees and plants and animals and tough protecting covers of trees, plants and animals are produced by the cell colonies for their use and purpose, to protect themselves, in precisely the same manner as man produces his houses, clothing, etc., to resist and protect himself from enemies and the elements.

The intelligence of man is inferior to that of the cell because he must learn and get his advice from the cell. We can see that the cell understands how to handle the different forces in nature like heat, light, sound, electricity, motion, etc. They use these forces for this and that purpose to produce certain desired results just as man does. They make and use heat to warm up their habitations. They make a light to guide their actions, and electricity for both offensive and defensive purposes. We find the cell employing all these blind forces also before he discovered the advantages of social life and while he was living the life of the savage. We see the cell living the single separate life of the savage today and we see him make darts and arrows. We see him go hunting for his food. We see him begin the social life as in the Volvox and sponges. We see him develop social habits. We see him finally in the highly organized colonies of cells, such as plants and animals. We see him in this highly organized social state build antlers on the deer. We see him build the fly traps on plants. We see him tear down and dismantle the caterpillar and with the same material build a flying machine we call butterfly, and we see him paint the wings and body of his flying machine, so as to be able to deceive all his enemies. We see him do all these things and we are forced to the conclusion that if man is intelligent so is the cell. We are forced to the conclusion that if man produced his various structures by reason of intellect, then the cell also produced his analagous structures by reason of intellect. There can be no distinction.

CHAPTER 10.

CONCLUSIONS.

We have now investigated and discussed so many points tending to show what plants and animals really are, that I shall close with this chapter, and state what I think should be the natural conclusions from the facts at hand.

We shall take a general survey of the whole situation again, and see if we are not forced to admit that all plants and animals are built by and for the cells, in precisely the same manner that the structures produced by man are also produced by and for men; that the cause of man's ability to produce houses, railroads, ships and other structures is the fact that he is intelligent; that the cell also is so in no less degree. In fact I believe that the cells living singly, like the cells that make weapons to kill their prey at a distance, and many social cells like those who build climbing plants, and also those who build and invent contrivances with which to catch and poison large animals like deer, camels and even lions, have a keener intellect than man.

We remember how Dixon showed that plant building cells must have not only all of the five senses that we have but also two more which he called psychic and physical sense, because it is a fact that some climbing plants will creep towards the nearest support, and if the support is shifted several feet from its former position, the vine will within a few hours change its course, in the new direction. This it will do even if the view of the support is obscured and concealed by ridges of earth.

How do we account for the "homing instinct" in the pigeons and the bee? We say we do not understand it; we give it a name, call it instinct, act wise, and say that those instinctive acts were at one time performed with a conscious intelligence and were repeated until they became automatic and instinctive. This method of explaining something not understood by giving it a name, leads only to confusion. If these plants and pigeons have a sense and intellect that we do not have nor do not understand, why not admit it? If instinctive acts are intelligent,—which fact they now admit,—why do they not admit also that they are intelligent whether produced by plant, animal or man?

We have not the ability to penetrate into the future and the unknown, but we should apply common sense to what we do know. The idea that there is nothing in life but matter and force I do not believe has been proven, but is yet apparently without foundation. We find matter and force in the organic world; they follow fixed laws. In the organic or living world, we find matter, force and intelligence. If the living things defy the natural forces, and act differently and only for themselves, we must admit the fact that life or living things are directed by an intelligent force. As far as we have explored the actions of natural forces from the molecule to the distant solar systems, matter and force obey fixed laws. In life we find all living beings from the cell to man moving contrary to the fixed laws of matter and force. They control and direct matter and force for their own purpose. The fixed laws of gravitation cause all water in streams to run into the ocean. The intelligence of man and the beaver can stop and compel the water to wait, and to serve man or beaver in building houses, dams or irrigation works.

Some people claim for the cell only irritability and not intelligence. The word irritability is meaningless. They poke a needle into a cell, or pour acid on him and when he jumps around trying to escape or defend himself, they say he is irritable. Such experiments and expressions are pure nonsense. You tie any man down, and poke his naked body with sharp iron bars, and burn him with acids and poisons and his actions would also be irritable, whatever may be understood by that word.

The actions of the cell when examined in his natural habitat show the same intelligence and foresight as man. Think of the amoeba, carrying with him material with which he can make a coat of armor and cover himself when necessary for self preservation. Whether this animal we call cell or amoeba came to exist by a chance assemblage of matter or not we do not yet know. It took place ages ago, before any plants or animals existed. We find the same struggle for existence down on the lower plane of life in the microscopic world where the cells live as we do in the life of the larger beings of plants and animals.

It is not necessary in discussing the question at hand to know how the cell came into existence. We are concerned at this time with the questions, who produces plants and animals, and how is it done. From careful investigation of all the facts, it must be answered that the cell produces plants and animals precisely in the same manner as man produces houses, ships and railroads.

The magnitude of the work done by the cell in the past ages as evidenced by the coal deposits and fossil remains, shows the enormous length of time in which he has been engaged in battle with the elements in his struggle for existence. The cells like other animals are of different sizes; they live on and off each other in the same manner as plants, animals and man do; the larger devour the smaller. The amoeba, which most resembles the man building cell, is among the largest. The organic material we need for food and clothing is manufactured by the cells in plants and animals from other raw material in the soil, water and air. The material generally used is water, salts, carbonic acid and ammonia. The plant cells can combine carbon, hydrogen, oxygen, and nitrogen to form these substances known as fats, sugar, starch and other carbohydrates.

The cell uses the heat of the sun to separate the oxygen from the carbon and the nitrogen from the hydrogen in the manufacture of sugar, oil, starch and other carbohydrates and which may be again reconverted into heat by burning in the animal fire or in the open fire. The only beings in the world that can and know how to do these things, how to manufacture these things we need, are the cells that build plants. Man is bending his energies in the effort to discover and grasp the secrets from the cell. Think what a change in the world when the secret is discovered, when man can change the crude elements into sugar, starch and oil for food.

We have made a beginning, as we have already discovered how to extract the nitrogen from the air and many other things made by the cell. You see it is the same with the cell as with man, if anything is to be done, the cell must do it. His intelligence must be there to guide and direct matter and force in such a way as to effect certain results.

If no intelligence is there to guide it, matter will always follow fixed laws. Air will stay air and water will remain water. The intelligence of man is necessary to control, guide, direct and arrange the blind forces and dead matter to effect the results desired, as in the manufacturing of nitrogen, or extracting it from the air. It is also equally necessary to have the same guiding intelligence in the production of this same material by the cell. If you furnish the hen building cells with lime, they will cover the eggs with a shell. The cell can not produce anything by magic any more than man. The material must be at hand, but the material alone will produce nothing. Some intelligent being must arrange or direct the arrangement of the raw materials, before we get the production of the structures.

The forces of magnetism, electricity, light, heat, and motion follow fixed laws. The cell takes advantage of these laws just as man does and by so doing produces the food and building material which he needs. The cell is no chance product, he is a living conscious being. Watch him under the microscope; watch his movements in the water, how he darts about in search for food, and how he regulates his speed, so as to avoid colliding with other cells; watch him as he chooses this food and rejects the other, and you will be forced to admit that his actions are voluntary and those of an intelligent being.

The cell has the same work to perform as man. He must provide himself with food and cover. He must take in food and dispose of waste matter. He must exercise all the bodily functions of man, such as respiration, circulation, movement and excretion.

We delude ourselves with the idea that intelligence can only exist where there is a brain and nervous system. We must remember that man, animal or plant is only a colony of cells and that plants with neither brain, nerves nor muscles can think and invent traps with which to catch insects, showing the most extraordinary skill and inventive genius.

As soon as the insect gets far enough inside the trap, the signal to slam the trap shut is given and the insect is caught. The plant cell then cuts him up into microscopic particles with an acid combination called gastric juice and the cells eat the insect.

In the plant called Sundew, the trap is also set and baited. The tentacles of this plant all work together, pushing the helpless insect towards the center where it is treated in a similar manner. It is digested with a liquid similar to the gastric juice of man. If anything touches those traps which is not good to eat, they get rid of it in the best way they can and pay no attention to it.

These actions are in every detail the same as those of intelligent man would be under similar circumstances. The individual directing the actions of that part of the plant composing the fly-trap show the highest degree of skill. The fly-trap is no chance product. It is a carefully considered, well-ordered scheme and arrangement, calculated and designed to effect a certain purpose. The forces at work here catching the foxy insects are no blind chance forces, they are intelligently controlled and directed.

We have investigated the effects of all the material forces, such as gravity, electricity, chemical action, light, heat, cold, wind, water, etc., and we have found that life is not any of these forces, but that the cell is affected by these forces in precisely the same manner as man, animals or plants. It is of course only what we should expect, as all plants and animals are colonies of cells, and the cell could not be affected by these forces any differently in one place than in another.

We have also found that the cell uses the raw material and natural forces to build and produce his structures, fight his enemies and assist him in his struggle for existence, in precisely the same manner as plants, animals and man do. This too, of course, is what we should expect to find, when we consider what the cell is and that he is the one who made us and directs our actions. His desires and needs must be, and are the same in one place as in another. Life, we see, is produced by the activities of the cell.

Consider how the cells live a separate parasitic life in the sex organs of plants and animals, taking no part in the general work of the body. They live a college life, taking a preparatory course, and in the meantime they are fed and taken care of by the other cells of the body. The cell performs all the work done in the body. The muscle cells do the hard work of pulling. The gland cells produce the different kinds of secretions, like milk, tears, oil, saliva, digestive juices, etc. The white cells, blood cells, lymph cells bring and carry away material, and the brain cells and nerve cells do the thinking and directing. The intelligence in man, as well as in the cell, is necessary to direct matter and natural forces. The whole fabric of man's civilization, as well as that of the cell colonies we call plants and animals, would collapse if this guiding intelligence was destroyed.

We should also remember that the cell is a highly organized individual, having all the special organs required to perform all the functions necessary to the life of plants and animals, and having in addition other special organs, the use of which is not yet understood by man. It is very likely that the cell has some special sense organs which give him a different and a superior intellect to our own. This is evidenced by some animals, insects and birds possessing special powers, like for instance the "homing instinct" in pigeons, which enables them to find their home under such circumstances as would seem impossible to a human being.

We know that some cells have a special machine and apparatus with which they can make their own food from the crude material of earth, air and water, like fats, starches and sugar. We can see the machine and see it produce the starch, but we can not see how they do it. We know that they use the heat from the sun as a force in the same manner as we use coal in the forges and furnaces of our factories. We use heat to create a molecular disturbance in matter, just as they do, but on a much larger scale.

All the structures made by cells, just as those made by man, show a purpose, intention and design. The bird is made to fly, the deer to run, the fish to swim. The feet of land animals are specially constructed to suit requirements.

Some evolutionists claim that environments produce organs; that cold climate will grow hair; that the particles flying in the air will produce a nose; that the vibration of air will build the ear, and the vibration of light will make the eye. In like manner, they claim that the ground produces the organs of walking, the air, the wings, and the water, the fins on the fish. Now this idea is absurd; environment will produce nothing; if anything is to be produced the cell must produce it. This can be clearly demonstrated by certain experiments heretofore referred to. The hair on an animal will grow when it is necessary to protect the animal from the weather. If the animal in a cold climate is kept indoors or covered with clothes the hair will not grow. It is the same with the callous on your hands or feet. You or the cells in you must provide something on the bottom of your feet or on your hands, to prevent the wearing into the flesh. If you protect the bottom of your feet with sole leather or your hands with gloves, no callous to speak of will grow. The intelligence of man or cell will produce a protecting cover for your feet or hands, but the ground will not do it.

It is clear that bees, ants, spiders, birds and animals can no more build their houses, provide their food and carry on their various domestic activities without conscious intelligence than man can build his houses and provide his food without intelligence. The actions of the plant catching an insect and the actions of man catching a rat are identical. The actions in either case are directed by an intelligent designer and builder.

A plant has neither nerves, muscles or brain, as we understand those terms; it has, however, the same beings, the cells. The cell is the builder and caretaker, and we find the same purposive and intelligent acts in a plant as we do in an animal. The old notion that only man acts from reason, and that other animals act without reason, is simply foolish. They prove that instinct has arisen from conscious intelligence, by the fact that the acts of walking and playing a piano, etc., are first directed by our conscious intelligence, and after they are learned, they become, as they say, automatic, reflex or instinctive, or habitual. Let them call it what they please, the fact is that our hands and limbs are trained and instructed, and as soon as the muscle cells and nerve cells in charge learn what there is to do, they take care of the work, permitting the cells in charge of our sense organs to do something else more important.

Every cell of your body has the power to remember and learn, just as the cell in your brain. The honey comb of the bee, the dwellings of the ants, and the nests of the birds are marvelous works of art. They admit that these structures are wonders in design and architecture, but that the beings producing them have no intelligence now, but that at some time in the distant past, these acts were performed with intelligence so long that they finally became a habit, automatic and instinctive. Can anyone comprehend how or when any animal, including man, was any more intelligent than he is today. It is nonsense. To say that a bird or an animal performs intelligent purposive acts blindly, is only playing with words. The acts of a being performed with precision and certainty to effect a certain purpose, are not done accidentally nor blindly, any more in the case of one kind of being than in another. When the insect, fly or beetle, gets inside of the trap of the plant called Venus fly-trap, the trap snaps shut and holds him. The plant or rather the cells in charge of the trap are notified that he is far enough inside of the trap, and so they slam it shut; as soon as the insect is caught, gastric juice is made to flow over him and digest him; when the insect is eaten, the trap is again opened and set for a new victim. If this plant is fooled by a substance not food, it will not catch it, nor will it try to digest it. The cells in the plant can distinguish and choose what they want. Can intelligent man build and operate a trap any better, or show any more intelligence?

The structure produced, occupied and operated by these cells has no blood, bones, nerves or brain like man, but the beings who direct the actions of this structure are the same and can not be distinguished from the others in general appearance under the microscope. If the cells of man and the cells of a plant look alike and act alike, why should one be any more intelligent than the other. The cell builds everything just as a highly skilled man or an architect does.

The mind which controls the actions of insects, ants and animals, including man, is the mind of the cell. It is this master mind that directs all living things, plants or animals; therefore, we find about the same degree of intelligence in all these structures. If you place mustard or acid on the thigh of a headless frog, it will scrape it off. It will try with one foot first, and should that foot be removed or hindered, it will attempt to scrape it off with the other foot. This simply shows that the brain or head is not necessary in order to give intelligent directions and orders to the limbs, and that intelligence resides in every cell. The digestive system, the circulatory system, the respiratory system and the lymphatic system are all special departments looked after in our bodies by cells, who have each their special and specific work to do. All such work requires the conscious intelligent directing of some cell. The star-fish, for instance, has no brain, but his acts are as intelligent as any other animal's, considering his habitat, environment and place in life. In fact voluntary movements take place in plants and animals whether there is any brain or not. It is very clear that no brain is necessary.

The cells, plants and animals, manage each their own affairs, in spite of the interference of the elements. They employ matter and force in building up their different structures, and they control and direct the force and matter employed. They select some elements and reject others; they override and suppress some physical forces and employ others, as circumstances require. A plant will grow upward in spite of the force of gravitation, if it is necessary that it get the aid of sunlight. The forces of attraction, repulsion, capillarity, adhesion, etc., are taken advantage of and employed freely by cell, plant, animal or man. The fossil remains of plants and animals, like the ruins of old cities, dwellings and tools once used by man, are a record and history of the past ages where each one can read and study for himself what has taken place in the past history of life. These ruins, however, are in no wise as interesting as the living beings which have left no ruins, like insect life and others.

Let us consider for a moment the subject of protective coloring. Some caterpillars not only color themselves the color of the objects on which they rest, but in many cases they change their shape so as to simulate some other object, like leaves, twigs, etc. This illustration by Mr. Schulte (Fig. 50) is a good example of what they can do. You notice in this picture how closely the caterpillar is able to imitate a young twig, both in color, size and general appearance. Man has just begun to understand the great advantage of using these tricks of protective coloring to deceive his enemies in battle. Colors intended to deceive the enemy are now being used on battleships, soldiers and artillery. The cell which builds and directs this caterpillar and the cell which builds and directs the actions of man, cannot be distinguished, one from the other, under the most powerful microscope. There can be no difference between the purpose, wisdom and foresight exhibited by the actions of this caterpillar and the purpose, wisdom and foresight exhibited by the actions of man, when he covers his body or machine gun to deceive the enemy. The actors look alike and their actions are alike. Plants as well as animals take stock of and comprehend their situation and environments. In this respect there is no difference in cells, plants or animals, each will change and modify itself to surroundings.

Some claim that life is caused by chemical action, that life is purely a chemical action of matter. It is useless to guess as to the first cause of life. We are concerned first about the cause of plants and animals, we must first understand plainly the cause of them and then it will be time to begin to understand the cause of the cell. In order to understand who builds skyscrapers and railroads, it is not necessary to understand who built man. The fact which I wish to make clear in this book is that the cell builds all living things we see, and the reason that he is able to do so is the fact that he is an intelligent being. The cell is a self directed, self acting, intelligent and aggressive being, superior to his surroundings and able to direct and control matter and the forces of nature just as man can. We find him so at this time; what he has been or was in the beginning, millions of years ago, we do not yet know or understand, perhaps some day we shall.

The idea that climate will produce fat and fur on animals in the frozen North, is foolish. These things are produced for a purpose by the beings or cells which built the animals. Water will not produce the swimming apparatus on a fish any more than it will produce a propeller on a steamship. Back of all these structures, there must be some one who can direct and form blind forces and crude matter to effect the desired purpose. In walking, my acts can be directed either by my brain cells or by the cells in my spine. In either case the actions are intelligent. The movements, whether directed by the one or the other, require the conscious intelligence of some one. This some one must be somewhere.

A great number of educated people of late have come to the conclusion that nearly all animals reason. If they would investigate and stop to think, they would soon be convinced that not only do animals reason but also plants because they are both produced and occupied by the same beings. The idea that some actions are reflex, automatic and instinctive because performed by animals is without foundation. There are hundreds of plants now known which catch and devour insects with as much skill and dexterity as man catches the larger animals. The common cell called the amoeba which lives in fresh water ponds, and which most closely resembles the man building cells, is no fool. He looks simple but he is not any more simple than a human being. He eats, drinks, breathes and moves from place to place. He knows what he wants to eat, because he chooses one kind of food and rejects the other. He carries with him a concealed coat or armor with which he can cover his body in case of drought or other danger, and which he removes when the water comes back or danger is removed. Where does man show any more mechanical skill or foresight than does this amoeba, the primitive ancestor of man?

We must not forget what the intellect in man is understood to be. Webster defines it as that "Faculty of the human soul or mind which receives or comprehends the ideas communicated to it by the senses." You will notice from this that the intellect is considered to be those groups of cells connected with your sense organs which receive the information and do the thinking. The five senses are the instruments by which the brain cells obtain information from the outside world, they have nothing to do with the thinking. The cells inside of the man do the thinking, and not the man, just as it is the man inside of the ship or submarine that thinks and directs its actions and not the ship or the submarine. We speak of the ship "Maine" as an individual, doing this or that, forgetting that it is not the ship which performs the various actions, but that it is the men who are inside of the ship. The same is true of the plant catching a fly; it is not the plant but the cells who are in charge of the plant that capture the fly.

When I speak of the cell I generally mean the largest cells and the social cells which have the habit of building themselves into social colonies, like plants and animals. The different sizes and species of cells are simply unlimited. These smaller cells called germs and bacteria seem to make food for the larger, in the same manner as animals feed one upon the other. It is evident that uncounted numbers of ultra microscopic species of germs exist that never will be seen. Those large cells that are well known, however, such as build for themselves coverings of a pearly, horny or flinty material, and still retain their separate single existence, as well as those that organize themselves into social colonies, like plants and animals, certainly exhibit all the characters of an intelligent being; they discharge every function of the higher animals. There is no doubt in my mind that every cell is composed of still smaller cells, and that the size of every cell depends on the number and organization of the primordial beings which together make up the cell. Every cell is a crowd of thousands or more individuals, each one having this special work to do just as the organs of an individual have their special work to do in different places of the body.

When the cell multiplies by division, there is simply an even division of the individuals, each half receiving his share of the beings and equipment possessed by the cell. Who is this being in the cell who takes charge of and directs the work of equal division of the cell? We see him do the work and we call him centrosome. Intelligence is clearly shown by this individual, who looks after this important work. A great number of the cells who build plants and animals live a single and separate life several months before they begin building the plants and animals. Such is the case with several sea and water animals and plants. After swimming around for some months, they finally decide to build and settle down in permanent habitations, like plants and stationary animals in the bottom of the sea. They anchor their bodies and colonies to solid objects; they move about and have organs of locomotion like other animals. The movements of the cell are never without a purpose, any more than are the movements of other animals. The cells dart about with both speed and skill in pursuit of other cells and other objects for food. They are careful to avoid each other, unless they attack and seize each other as prey for food. Their movements are clearly voluntary and intelligent. Some have paddles, some have tails and some have other mechanisms for propulsion; their methods vary just as methods vary among animals.

When we comprehend correctly what life is, that all plants and animals are the productions of an invisible intelligent being, it is then easy to understand it. It is impossible for us to do the work of the cell just as it is impossible for the cell to do our work. We can, however, understand what the mind of the cell is, from our own mind. The cell has the same problems to solve as man. Matter and force are the same everywhere in the universe; they affect the cell as they do man. To direct the blind forces and to form crude matter so as to best subserve his purpose, are the problems of the cell as well as those of man. The amoeba must provide himself with food, and have material with him ready at hand at all times, with which he may cover his nakedness when necessary. The power to select and carry the raw material that will instantly harden into a flinty armour when spread around his body is a great discovery and a knowledge which must be passed on from one generation to the other.

Certain physicists, chemists and physiologists claim of course that there is no such thing as life or intelligence in the universe; that there is nothing but matter and force; that force goes always with matter; that matter under certain conditions could and would assume life; and that no intelligence is required to guide matter. It is, in my opinion, useless to make any such statements at this time. Whether matter could have got started some time in the past million years, and developed finally into the intelligent being we call the cell, which builds and directs the actions of man, we do not know. But from the evidence at hand it would appear that life, the intelligent force, as distinguished from the blind forces of nature, is a separate force in the universe. A clear line can be drawn between the dead and the living. Intelligence must have existed before structure. The idea must first exist how to build and what to build. From the facts before us it would appear that a separate intelligent force may exist unseen in the universe, separate from matter. The living intelligent cells, like man, control and direct matter and are masters of it. The cell that builds man, and the cell that builds the tree that lives a thousand years, look alike and are alike; the only difference in them is their record of past events, their experience and knowledge. This same is true of animals and man. The general appearance of a man who knows how to build a skyscraper and that of the man who knows how to build a hut, is the same; the only difference in them is in their knowledge,—in the experiences stored up in their memories. Every species of cell can only know how to build the structure from which he came.

The fact that everything is first produced before it is used shows clearly that some one is building for a purpose. At the right time milk will be made ready as food for the infant. Why is it always produced at the right time and always when needed? The sense organs in the same way are prepared and ready in advance, before used, and are made for a purpose; they are not chance productions, but are in every way similar to the intelligent purposive acts of man. The fly-traps of plants and the electric organs in fishes are in the same manner made for a use and purpose, in accordance with the ideas in the minds of the makers. Not only do we find organs made for a specific use but we also find that they are in charge of beings that know how to use them.

Plants hibernate in the winter, just as many of the animals do. The cells of the plant build the leaves whereby they are able to get in touch with the sunlight and make their food and other material in the summer; when fall comes they go back into the tree and sever the leaves from the tree. The twining plants produce first grasping tackle they call tendrils; the plant then moves its head in a circle, feeling in that way for quite a distance around itself for something to grasp. The builders make first an apparatus with which to fasten their structure onto other objects. They know what they intend to do and at the first opportunity they do it. The plant cells show also great intelligence and skill in producing brilliant colors and sweet perfumes, to lure insects and compel them to serve them in numerous ways.

The limited ability of the human mind to comprehend things, and its liability to misunderstand things,—even of the minds of those who should be the most likely to see things in the right light, are clearly shown by Mr. Loeb of the Rockefeller Institute in his description and interpretation of the actions of the young beetle, which came out of the ground in the spring hungry and knew how and where to find something to eat; without any previous experience it went directly to the top of the tree and ate the buds from the end of the twig. The ability of the beetle to go directly to the top of the tree where he could find something to eat, without having had any previous instructions or experience, Mr. Loeb calls "heliotropism" while others call it "instinct." There are two kinds of beetles that I am very well acquainted with in this country, that live on the outside of the bark on trees and they are a source of considerable amusement. You cannot possibly see them except on the instant when they move. They have their backs shaped like the rough bumps on the bark and of course also have the exact color of the bark. I can generally find one by passing my hands or something over the bark of trees, especially of oaks. When disturbed they will make a swift rush for a new place to hide, and it will always be in a spot where their shape will exactly simulate the edge of the bark. When I discover one I usually call someone's attention to the spot and ask him whether or not he can see the beetle, and never yet has anyone been able to see him even when I pointed directly at him, until the beetle made a swift spurt to a new location. When I read Mr. Loeb's idea of what was the cause of the young beetle going directly to the top of the tree to eat the buds, I thought to myself here absurdity has reached its limit. It seemed to me that when a man like Mr. Loeb will believe and will attempt to make others believe that it is the sunlight that makes the beetle go to the top of the tree and eat the buds, that it was time for someone to try to get the tangled mass of contradictions and mystic absurdities straightened out so that a person with ordinary common sense could understand it. The idea seems ridiculous to me that a beetle should be heliotropic because he goes after something to eat, and especially that as soon as he has eaten a bud his heliotropism has left him and the sunlight no longer attracts him. He goes even so far as to state that it is a lucky thing for the beetle that the heliotropism leaves him as soon as he has eaten the bud, for if it did not, the beetle would be stuck fast, helpless, at the top of the twig by the upward pull of the sunlight, which he calls heliotropism. What has become of our thinkers when such statements will stand unchallenged and undisputed?

Mr. Loeb as well as others should know that sunlight has no more effect on the beetle to pull him in any direction than it has on other animals; and he should know that the sunlight or heliotropism will not attract a beetle towards his food any more than the same force will attract a cow towards a haystack. He should know that the beetle prefers darkness, and goes directly to the top of the tree for something to eat in the darkest nights, as well as at other times when he happens to get out of his house in the ground, where he has been rebuilt from a worm into this beetle by the cells that occupy him. The cells that build the tree beetle need not "take a back seat" for any other cells in the world for inventive ability and constructive skill, not even for the cells that build and guide Mr. Loeb or Mr. Rockefeller. I think the tree beetles exhibit a most wonderful skill and intellect in the builders, as is shown by the many inventions of shapes and forms, in these illustrations (Fig. 51) from Mr. Schute's Evolution. A tree beetle is a habitation built by a colony of cells, so as to be able to dig holes in the ground, run up and down the side of trees and navigate the air; and he is of such shape and color as to deceive all his enemies that might wish to devour him, and he certainly is a great success. He is with us today while the larger extinct structures are with us only as fossil bones and remains, and he will be with us on this earth as

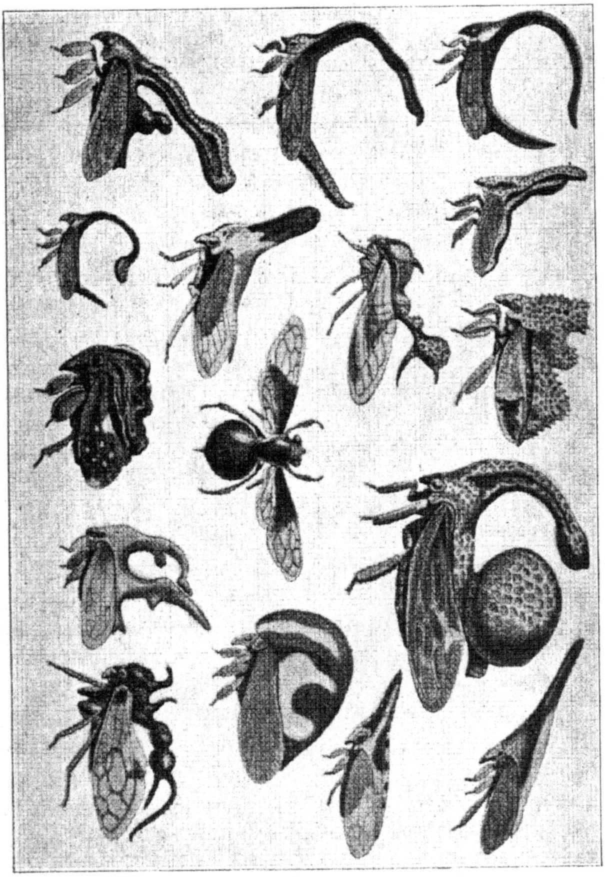

FIG. 50.—Central American Leaf Hoppers resembling the prickly and thorny growths of plants on which they presumably live. Certain of them also represent gall growths on the plants. Protective resemblance. [Figures collected by Dr. L. O. Howard, from various plates published in the Biologia Centrali-Americana.]—SCHUTE.

long as man, and when a great number of larger, and what we consider more intelligent individuals or cell colonies are no more. The cells that made and guide the beetle can clearly remember from their former experience in the parent beetle where buds could be found in the early spring, and so they guide him directly to the place in the top of the tree. In precisely the same manner would an experienced captain of the submarine or ship guide his structure directly to the proper place, by reason of his knowledge and past experiences. There can be no mystery, instinct or heliotropism about it.

The plant called the Venus fly trap with its actions so similar to man's in making traps and catching flies, is a very good illustration of design, intellect and foresight by plant building cells. Still it is no better illustration than many others are except that the actions of this plant resemble so closely those of man. Take for instance the fly traps made by the plant called the Darlingtonia. This trap is a hood or box into which the insect is decoyed, and the hood or box has transparent patches just like glass, which lead the insect to think that they are openings, and the insect attempts to escape through these windows just as a fly attempts to escape through a pane of glass. We could make very effective fly traps in the same way,—in fact, we use some now made on the same plan. The mind required to conceive, invent, and construct these traps with these window lights for the purpose of fooling the insects and in that way capturing them for food is in no manner different from our own.

A plant called utricularia growing in the bottom of ponds can uproot itself and can produce a gas, by means of which like a balloon it can come to the surface and flower. As soon as it is through flowering, it again becomes heavy by releasing its gases and settles to the bottom of the pond, where it fastens itself again with roots. Consider the elaborate machinery used in these operations and the skill required to operate it. The plant roots and uproots itself at pleasure; it makes a gas with which to raise itself to the surface just when required. The individuals who build and operate this plant must necessarily understand and have as complete a knowledge of hydrostatics and mechanical designs as the builders and operators of our submarines. The ability to make, keep and release certain gases at just the right time so as to cause the plant to float or sink as required, shows precisely the same skill in this plant building cell as in man. Before it is ready it is anchored by its roots to the bottom; as soon as it is ready to float, they cut the roots and release themselves. In what manner do these various actions differ from those of man with his air ships and submarines?

Every flower has some peculiar scheme of its own to compel insects to brush up against the pollen and carry it on their backs or in some other manner around to other plants. It would require volumes to describe all the different schemes and ingenuity displayed by plants to cause dispersal of their young. The more you examine into their methods, schemes and inventions,—especially those of vegetable life,—the more sure and certain you become of the fact that the cells who build plants and animals and those cells that resemble these in size are all possessed of the same degree of intelligence. They all display the same conscious wisdom, foresight and inventive genius in their place in life as man does in his. The male salmon after having fought with other males and obtained possession of the female scoops out a hole in the sand where the female lays her eggs, and then he covers the nest again with sand to prevent anything eating his young. The young salmon is provided with food enough to last him until he can get his swimming apparatus and mouth ready, so as to be able to catch his own food. We find the same thing to be true in the case of man; the germ cell is also provided with food to last about five days or until it can be connected with the food supply from the body and attended by the body cells. You see in every place where you look into the actions and methods of the cells, whether in their stationary abodes, the plants, or in their movable dwellings we call animals, the same wisdom and foresight is displayed. Call it intelligence or what you will, it is the same as that of man. The building of the animal body takes place in exactly the same manner as the building of a large structure by man. Some cells work at this and some at that; some build the bones, some the limbs and some the heart and arteries, etc. They all work at the same time and in harmony with each other; each crowd does its part, expecting every other crowd to do its part too, so that when the structure is completed, the whole will work harmoniously together as one whole machine. They must understand each other and what they are about; and while the builders are rushing the work, the body cells must do their part and furnish the food and building material in abundance. When the time comes that the new structure must be taken care of in the outside world, the food is still furnished by the cells on the inside, as they know best what he needs, and the food is then supplied to the new structure at a different and more convenient place outside.

The plant furnishes every young plant or seed with enough provisions to give it a start in life in just the same manner as animals. Every kernel of grain is the cell provided with food to give it a start in life, and we crush and eat both the cell and his food. It has been clearly demonstrated of late that the cells at the end of the roots exercise the same selective power as the cells in the stomach. Knowing what is required by orders coming from different places of the body, they select those particles and material which are called for, and reject all others. The materials selected by the cell at the surface of the roots are carried inward and up to the places where they are required, in the same manner as in animals; and the leaves absorb all the moisture they can from the air, and what more is necessary is brought up from the roots. The plants have in a way a mouth at each end. They have a circulation of the sap in different directions, up and down, for the purpose of transportation, just as animals have a circulation of the blood.

We do not know very much about the beings that go to make up the intelligent individuals we call the cells. We see them move about in the cells, produce this and that, and we call them granules, just as a number of years ago, we gave the name cell to the intelligent builder of all life on earth that we now see, and which we call plants and animals. When the cell was first discovered it was thought to be merely a crystal of matter and was not considered a living individual. Man could not then comprehend that he had discovered his maker.

It is singular how the cells of plants are affected by chloroform, drugs and poisons in precisely the same manner as man and animals are. It can be observed in sensitive plants that they cease to be sensitive when chloroformed; and sensitive plants who close their leaves when jarred or disturbed will get used to it if this disturbance is kept up for a time and will pay no attention to it, being affected in the same manner as man and animals would be. There is a plant called Nepenthes, which has an insect trap like other pitcher plants that catch insects, but in addition to the ordinary method it makes a poisonous gas heavier than air which floats over the water in the pitcher. The insect is attracted and fooled into this gas by perfume and honey and when it becomes poisoned and suffocated by this gas, it falls into the fluid below. This fluid is possessed of digestive properties similar to those of gastric juice, and the cells of the plant digest and eat the insects. Where does man display any more inventive genius and intelligent schemes, by which to catch animals for food? Do you suppose that the manufacture of this heavier-than-air poisonous gas could come about by chance? While every plant and animal displays actions purposive and intentional to affect some future end, still the insect catching plants are the clearest illustrations of intelligent acts because they are analogous to those of man. It is a singular thing and the cause is not yet understood, why only the meat of insects and particles good for food become stuck and fastened in the sticky secretions on the leaves of insect catching plants. The sticky secretion is in composition similar to that used by man to catch flies, but it is so manipulated that if anything like leaves, gravel, etc. touches it, it does not stick but for an instant, unless good to eat. How are the cells of the plant able to manipulate the stickiness of the fluid so as to stick in one case and not in the other? The human mind with its limited capacity has not yet been able to understand or comprehend this secret of the plant cells. The plant has specialized organs just as the animals have. When the insect is caught, one crowd of cells holds him with the trap while another bunch pours digestive fluid over him and the third group carries his several microscopic parts to the interior of the plant where he is eaten by the whole colony of cells that make up the individual we call insect catching plant. The seizure, capture and digestion of insects by plants require precisely the same intelligence as do the capture and eating of an animal by man. There is no difference in the general appearance of the plant building cells and the man building cells. The only difference in them is their general knowledge, experience and education recorded in the memory of the individuals. In the past ages the cells have each accumulated the special experience and knowledge which now guide and direct their actions in the same manner that the experience and knowledge of every individual determine and guide his actions. In the insectivorous plant called the Sundew, the tentacles with the sticky material gradually push the insect over to the center of the leaf where he is covered with a digestive juice similar to gastric juice and digested. It is observed that the juice begins to flow into the center of the leaf where the plant expects to land the insect as soon as he is fairly well secured. Intelligent man will act in a similar manner. As soon as the animal is caught those who have charge of preparing him for food start their work, anticipating and knowing that the others in due time will land him in the camp, ready for the kettle.

The plant is not fooled by other objects falling on the trap and the trap is again set for another insect as soon as the entrapped one is digested. This is precisely what man would do under a similar circumstance. The amount of digestive fluid produced will always be in proportion to the size of the insect and no digestive fluid will be wasted. If the trap is fooled into making a closing move by some other material than food, no digestive juice will be produced and it will very quickly discover the deception played on it by man and again open its trap. It behaves exactly as man would under a similar circumstance. The individuals of this colony of cells known as the Venus fly-trap, and others, understand how to make an elaborate fly tray and into this they entice and capture insects for food. They are not in need of this food because they understand how to make food from the raw material of earth, air and water by the aid of sunlight in the same manner as other plant cells. The fly trap could not have arisen or come to be by any beneficial variation because it could be of no benefit until completed as a fly trap, and the catching of flies and insects is a mere sport and not a necessity. The whole scheme must have been conceived, planned and discussed and experimented with in this cell colony for a long time before they got it into a working condition. No chance variation known as beneficial variation could have possibly produced it. The idea of catching insects must have existed in the minds of the builders before they began to prepare and construct these fly traps. The idea must come before the structure in all productions of art by man, and it must have been the same in this case with the cells producing the fly traps. The methods used by other moving and sensitive plants like those that sleep and climb, are similar to the methods used by animals. The climbing plants attach themselves by grasping hooks and suckers, and direct the several movements by which they feel around for supports with a purpose and end in view similar to other animals; with the aim in view of climbing up into the sunlight they take advantage of the principles of engineering. The schemes of plants for catching insects, however, are more readily comprehended and recognized by man as intelligent because they are similar to his own. The tricks of the professional trapper and hunter are recognized in the actions of the insect catching plants and these actions show an intellect and skill similar to those of the trapper. The acts of capturing insects are the same as those of the hunter, who sets out his decoy birds and lies in wait for the unsuspecting victims.

It seems clear that in order to produce the organs of our body, like the eye for instance, the cells must have a mind similar to that of the man who makes optical instruments; they must be fully acquainted with the law of optics. Chance or natural selection knows nothing of optical law and cannot produce the eye any more than it can the optical instruments. The same may also be said of the other organs and especially of the powerful electric organs in fishes. While these electric organs were merely in their experimental state, they could not have been of any advantage to the fish in his pursuit and capture of other animals for food. The inventive mind of the builders of the fish must have been the cause in this, as in other inventions; accident and chance could not produce electric storage batteries and apparatus with which to gather and discharge powerful currents of electricity at will. Accident and chance could no more have produced them in the fish than in the power plants we now have in our cities. The fish is a colony of cells and the city is a colony of men. Why should the production of the cell be that of accident and chance, and the production of man that of intelligence? Man attempts to argue that the actions of other animals are not intelligent and calls these actions automatic, reflex, habitual and instinctive. When confronted with the fact that these instinctive and automatic actions are as intelligent and purposive as his own, he makes the absurd statement that they were at some prior time directed by the intelligence of the animal until they became habitual, automatic and instinctive. When could man or animal have been any more intelligent in the past than he is today? The words, "variation," "natural selection," and "adaptation" used by many evolutionists as a cause of the production ofdifferent animals and organs of animals are a delusion,—they are only a playing with words. The multitudes of cells that make up the colony or cell republic which we call a plant or animal, must build or produce the structure, for the climate and chance variations will produce nothing. It is clear that in the battle for existence among several contending for space, the best fitted individual or battleship will generally win out and survive and perpetuate its kind. No one ever questioned or disputed this fact, but it is only a fact and law in the universe, and has nothing to do with the original construction of the battleship. The cause of the battleship was the intelligent being who produced it and directed its action in battle.

Many writers use words in such combinations that they become complex and misleading; some wish to appear wise and invent words and phrases to make you think they understand something which they do not. These people are an obstruction to a true understanding of life by the busy man. When Darwin, Wallace and Spencer wrote their theories of evolution, they did not comprehend that the cell was an intelligent living animal, so they could not understand the cause of the different structures in life. But they could clearly see that all plants and animals were not born or made alike; that all were a little different and that the best individual won out in the game of life, and again produced his kind.

We are told by Darwin, Haeckel and other evolutionists that accident and chance produce the living things we see, like plants and animals. Others tell us that an unknown Creator produces them. Both sides seem to grope about in the dark for a cause and deliberately ignore the plain and real cause. It is just as clear that the cells build plants and animals for their own use and purpose as is the fact that man builds houses and ships for his own use and purpose. One shows precisely the same intellect, wisdom and foresight in his place in life as does the other. They mold and form the crude matter and direct the blind forces in the universe for their own use. The cell is the builder and thinker in man as well as in all other living individuals. We know that the cause of man's ability to produce things is his intelligence, and we know that the intelligence of man arises from the thinking cells in his head, hence it seems to me there is no room for argument.

When we understand life correctly, that the cell is the cause and builder, that he is intelligent, that he has memory and knowledge, that he can only act from experience, just as we do, the mystery of heredity disappears and the inevitable development of a kernel of corn into a cornstalk and not into a sunflower becomes perfectly clear. These cells have never had any experience or knowledge of how to produce anything but the corn plant. For that reason, the corn stalk must necessarily be their specific production. The cell lives in a different place in life—in a different world, the microscopic world, but nevertheless we can tell by his actions what he is and what he does—"By their acts we shall know them." We know that the thinking cells in our brain have the power of memory. This gives us a fairly good understanding of what memory is in the cell. Why like will produce like and why, in many cases, like bees and ants and caterpillars and butterflies, they produce not only like but also entirely unlike descendants, is clearly understood because, like man, they produce what they consider is best and necessary, and what comes within their knowledge and experience.

We do not know the cause of chemical force, electricity or gravitation; we do not know the details in the construction of the cell, but it is not necessary to the question at hand, and when we do not know it is best and right to frankly admit it. In no place do we find intelligence in the actions of the natural forces and dead matter. We know that an amoeba when necessary can build around itself a hard protective cover, and to be able to do this, it would seem necessary that a number of separate individuals in the cell would be required to perform the act. Who the builders are and how they work in the amoeba we do not know because we cannot see them plainly enough. The natural forces will act whether anything is necessary or not. The actions of the blind forces like electricity, etc., take place regardless of the will and judgment of anyone. Hence this case of the amoeba putting on his coat only when it becomes necessary, it shows at once that intellect is in command, directing the natural forces and matter, and that the amoeba is an intelligent being. The centrosome or general manager in the cell, who gives orders to the rest of the primordial beings who make up the cell to either divide or to cover itself with a protecting covering, is without doubt an intelligent being, as appears from his actions, which now can be seen with a microscope. Not long ago the cell could barely be seen and we thought it only a crystallized matter of some sort. The cells build their structures for a certain purpose, expecting and intending that they shall be used for that purpose.

However, if it comes to pass that they will not be required, they will be gradually removed and abandoned. This we find to be the case with all organs and even the eye will lose its power of sight and will gradually disappear if not required,—as is the case with fishes which have acquired the habit of living in dark caves. This we find also to be the case with man in his use of structures and machines; whatever is found to be of no use or service would be abandoned and discarded as useless. The light does not produce eyes nor does the vibration of the air produce ears. The cell wherever you find him shows by his actions the same intention and purpose as man. In every place, the actions are intelligent and voluntary. The amoeba, the single cell, which lives in the water and so closely resembles the man building cell, can move in any direction it thinks best, eat what it thinks best and when in danger escape from its enemies, and, under exceptionally dangerous circumstances, it can cover itself with an extra tough and hard cover wherein it can remain until the danger is over. Other species of its kind make and wear permanent covers of pearl, horn or flint wherein they live and paddle themselves around through the water in search of food. Others not only cover themselves with armor but provide themselves with flash lights to better see and capture their food and some make darts with which they can strike and capture their prey at a distance. The other social cells that multiply and aggregate themselves into colonies and cell republics which we call plants and animals, show the same inventive skill as these single cells. Their special skill consists mainly in being able to produce large protective colonies.

It is hard for the ordinary man to comprehend that an animal or plant could have any intellect without a brain, head and nervous system. The easiest way to illustrate intellect in the plant is with the actions of the various insect catching plants. The fact that these fly traps are of the most scientific construction and are operated in precisely the same manner as if they were operated by man makes it clear that the cells who build and operate the traps possess at least the same skill as man. The old ideas about instinctive actions of birds and animals are all cleared away when you understand life correctly. How can a swallow fly the first time it makes the attempt without having ever had any practice or instruction in the art? It is clear that the cells who will do the flying are the same individuals who have made the swallow. They have designed and put him together, piece by piece, for a specific purpose, and every part of him was placed in the correct place and made of the right material. Every individual cell that makes up the swallow is in his place, and knows what he is there for and has charge of the particular business in which he is skilled. There is no difference whatever in the various purposive activities of the cells building a swallow and those of man building an air ship. Both are built for the same purpose and both fly in their first attempt without any previous experience or instruction. It is commonly said that the swallow flies from instinct, then why not say that the air ship also flies from instinct. Yet this would be ridiculous; every one knows that it is the builders who set these machines which they have produced into motion. There is no distinction, in fact, between intelligent and what they call reflex, automatic, habitual or instinctive acts. All these are intelligent, purposive acts.

There is, in fact, no difference between a voluntary and an involuntary act,—both are intelligent and purposive; for instance, I can wink my eye voluntarily and involuntarily. If voluntarily, the act is directed by my sense cells in charge of consciousness; if involuntarily, the act is directed by the nerve cell having that particular business in charge. Why is the same act any more conscious and intelligent if directed by one cell than by the other? If it is an intelligent act in one case, it must necessarily also be so in the other. It has been shown that in civilized man there is a bunch of cells near the side of the head specifically delegated to direct all the actions in writing. This "writing center", as it is called, is located near the speech center and if these cells who have the writing business in charge are injured the individual will not be able to write a word and all his knowledge of the art will be entirely lost. It is the same with any other centers or specifically trained cells in your head. If they are injured or destroyed, that particular work or faculty will thereby be lost to the individual. The brain is simply a bunch of centers, each center or bunch of cells having charge of their particular department of work. When you kill or remove those individuals, their work ceases with their destruction or removal. When we first learned to write or walk the sense cells had charge of it, but as we became more efficient other cells took charge and learned it and finally relieved the sense cells of the work, so that they could do something else.

When life is correctly understood, all these theories about irritability, excitability, emotion, automatism, reflex action, intuition, instinct, etc. become perfectly clear and it can be seen that they are all the same thing, and that they are all intelligent actions coming from differently situated cells. How the cell, matter and force came into the world we do not yet clearly know, but we do know, however, how plants and animals come into the world because we can see the actions of the builders that produce them, which we were not able to do some years ago when we were compelled to say that they simply grew. By virtue of his intelligence, man is able to control and direct matter and force so as to effect results such as houses, ships and cities, and in an analogous manner, we find the cell is able to control matter and force in the universe so as to cause the structures we call plants and animals. We can only deal with these things we see and know, such as cells, matter and force. In reference to those other questions that we do not know, such as how the cell came into the world or how force and matter came to exist, one has as much right to guess as the other and it does not prove anything. The power of memory is without doubt the cause of progress and intellect in the cell and no doubt some day this power will be better understood, and I am satisfied that it will be based on some physical and chemical skill understood by the cell. I feel satisfied that we shall find that in some perfectly natural way the record of past incidents and occurrences are recorded and preserved in the same manner as man now records and preserves his experiences by means of books, paper or phonograph. You might ask where would the cell, this microscopic individual, have room for any such record. It is possible for the cell to carry with him such records, for we find that the amoeba, which is the most similar to the man building cell, is able to carry with him material and force sufficient to build around himself, whenever so desired, a house or cover, solid as steel, wherein he can remain safe until the dangers which threatened his life are over. Looking at him with our most powerful microscope we can, as yet, only see what appears to be a mass of microscopic and active granules, running hither and thither. Where does the amoeba keep all these materials with which he can, when he thinks it is necessary, construct this coat of armor about himself and where does he keep all his specifically trained force, who understand how to put the raw material into a covering for his unprotected body? When we shall have produced a microscope powerful enough to enable us to see the detailed activities of the primordial beings that go to make up the individual we call cell, or the amoeba, then we shall very likely be able to discover how he is able to keep a record of past events and is able to refer to such records when necessary to guide his actions, which we call memory.

Without this power of memory, the cell could not build the plants and animals we see, nor could the civilization of man be maintained without his ability to keep a record of the past to guide his actions. The physical forces, provided with the same material to work on, always produce the same results; they always have done so and always will. The rivers always have and always will flow to the sea; the heat of the sun has moved wind and water in the past as it does today; the chemical actions and crystallizations have taken place in the past just as they do today and always will. Knowing the combination, the chemist knows the result. In life it is different because the cell is an animal or individual like man, who is able to override and direct the blind forces of nature to his own use. The environment of climate, wind, earth and water is made to serve him and does not control him. The young barnacle, called copepod, looks like a cross between a crab and a fish; he has a nervous system, eyes and limbs for locomotion. He swims about in the sea until he fastens onto a fish and there becomes a stationary individual. He tears down those structures and organs which are of no more use to him, like eyes and limbs, and becomes nothing but a sack fastened to the fish. This might seem a very foolish thing to do from our point of view,—that is, to destroy one's eyes and limbs and convert one's self into nothing more than a sack or stomach. However, looked at from the correct point of view, it is the right thing to do if one can do it. While a free swimming animal the copepod has to lead a very strenuous and uncertain life. While moving about in search of food, he is liable to be swallowed and his existence terminated any minute by hungry fish or other enemies. If he can securely fasten himself to a fish or some moving structure as a ship, for instance, he can ride around with plenty of food at hand and in comfort and safety for ever more. We must remember that this barnacle attached to some object is simply a colony of cells like any other plant or animal; and that the cells that made up his limbs and eyes and who were occupied with the strenuous work of locomotion and seeing, can now quit this work and ride in their little house in comfort and safety. How can environment, such as the climate or the water, cause or direct these actions and structures of the barnacle as claimed by some evolutionists? It is simply absurd and unthinkable. When the cells in the fish like barnacles make a sucker-like apparatus with which to effect a quick attachment to the fish should an opportunity arise, they must necessarily know what they are about as it is a difficult task not only to make the instrument with which to effect the attachment, but it is also a difficult task to approach and attach themselves to a fish. It could not come about by any accident and chance any more than a stone could jump and ride a wild horse by chance.

The acts of the cell whether in a barnacle, oyster, plant or animal show the same intelligence everywhere. They direct the actions and destinies of all individuals, just as man directs the actions and destinies of all his structures, such as submarines, battleships, etc. The cells direct the animals in their search for food; they direct the birds in their nest-building, and in their migration south upon the approach of winter; they direct the bee in his flight for honey, and the ant in his social economy; they direct insectivorous plants in their capture of insects and man in his civilization. The seed from the wild parsnip, which in its wild stage is tough and wiry, will in a few generations in a good garden soil, well tended, produce the plant as it now is. Why? Because in nature the parsnip has certain enemies and elements to struggle with which require that particular kind of tough condition of the plant in order to exist, and it is also precisely the same with animals and man. The wiry, tough wild boar must remain so in his natural habitat in order to be able to exist.

We see the same struggle for existence everywhere. It is the same among the single cells on sea and land as it is among plants and animals, and as it is among men in business, and among nations for a national existence; the same schemes are invented by plants to perpetuate their existence as by man, animals and nations. There is a plant in Persia that puts forth a pair of hooks of material strong as steel which fits over the noses of antelope, deer, and camels, and kills its victims, after capture, by the injection of a poison. It does this with the purpose and intention that its young plants shall feed on the decaying carcass while building new plants like those from which they came; this shows an inventive genius and foresight of the highest order. The idea that any chance variation could ever produce this far seeing invention to effect the purpose intended, is as impossible as that an automobile could be caused by chance. The cell which built this plant shows that the cause of evolution is the intellect of the builder and not chance variation.

If pigs are turned out into the timber and forced to protect themselves as best they can, they will slowly and surely revert back to the original type because it will be necessary for self protection. In order to be able to live on the scant supply of food in the timber, the pig must hunt and hustle all the time and sometimes fight for his life. He cannot have short legs and a heavy fat body and exist very long because he cannot get around quickly enough to gather food, even if he could protect himself against falling prey to the fox and the wolf. The builders are wise—they know these facts and as quickly as possible they will change their structure to a veritable fight-

FIG. 51.—Wild Boar contrasted with a modern Domesticated Pig. Reproduced from Romanes' "Darwin, and After Darwin." By courtesy of The Open Court Publishing Company.—SCHUTE.

ing and running machine, which can fight, dig and run with such speed and dexterity as will be necessary for existence. For this reason the neck and nose will be increased in size for fighting and rooting, and the rest of the body will be made to correspond to it and the legs will be made as long and strong as necessary to make up the perfect running, rooting and fighting machine which will be required in a rough and tumble existence with the elements and enemies that he will meet in the forest. The cell whose wisdom must look after these details for the hog will do the same for man, if he should be compelled to hustle for himself in a wild forest. Man would also be compelled to revert back to his savage ancestral type. The cells, like man, will build such structures as they know will be required in order to exist in the battle of life. Climate, earth and air, which are called environment, produce nothing, but they cause the cells to build structures differing in various places, depending on what is required. Intelligent man does the same and so do all living things.

Every part of your body, which is, in fact, a specialized colony of cells, does just what is required and just the right and proper thing in each particular case. Consider for a moment the complex and skilled actions of the cells that make up your mouth, tongue and throat, how they guide the direction of the food in one way and the direction of the air in the other; how dexterously the tongue is able to push the food around and between the teeth, hardly ever getting caught; how the muscles of your tongue and mouth form themselves into shapes to cause sounds of all kinds. Those actions are as truly intelligent as the actions of your hand in painting the most artistic picture, but no more so than the skillful acts of plants in catching insects. The ear is an instrument that, if kept up in proper condition, will detect the vibrations of the air, but in order to be able to hear the ear drum must always be strung up to a certain degree of tension; if not kept just so, you cannot hear. Who looks after these details? We find in every place in the body that every

FIG. 52.—Man and monkeys.—HAECKEL.

particular thing to be done and detail to be looked after is in charge of some cell or crowd of cells.

A large brain or head does not necessarily contain any more intelligence than a small one. Take, for instance, the brain of the ant, which contains but a few cells. Darwin himself states, "It is the most marvelous atom of matter in the world," and it has been clearly established that the ant is possessed of a very high degree of intelligence. For instance, all writers seem to agree that they store food for the winter, they live a social life like man; they build complicated nests and places to live in; they carry out the principle of division of labor; they adapt themselves to circumstances; they capture and employ slaves; they confine in captivity Aphides, insects which secrete a kind of honey which the ant uses as food. They have perfect sense organs and memory, so they can recognize and remember each other. They are emotional and display affection for their young, which they caress, tend and nurse. They can communicate with each other and in case of danger and difficulty, they consult each other and agree upon certain lines of action. It is also shown that ants gather the eggs of Aphides and treat them just like their own, guarding and tending them with the greatest care. The Aphides are the same to the ants as the milch cows are to man. And in a manner similar to that of man caring for his domestic animals, ants care for these Aphides. They even make covered ways between their houses and the trees and plants on which the Aphides live. They gather the eggs of the Aphides in the fall and protect them during the winter, then bring them back for hatching in the spring. They farm the Aphides in precisely the same manner as man takes care of his live stock. They also keep other insects, such as gall insects, caterpillars, etc., and in many cases, when necessary, they keep their insects in separate stalls. It is considered a common thing for one animal to capture another for food, but to capture another and keep him as a slave to do the work for him is another matter. We recognize our own mind in that of the ant. If a few cells in the brain of the ant show the same mind and intellect as our own, why is not every cell in the ant also intelligent? Why is not every cell in the ant's brain endowed with as much understanding as any other cell? Bees also are skilled workers—they reason in the same manner as man. If, at the beginning, their work has been started wrong, they will tear it all down and change the part so as to finally fit it to the general plan preconceived. They will sometimes modify the whole structure in order to meet and overcome some extraordinary difficulty to be encountered later on. I have not space to describe further the intelligence of bees and ants and other insects, but they have been lately studied and there is beginning to be a general agreement among most of the scientists that such insects as bees and ants show the same intellect as man possesses. The fact that the brain of the insect is microscopic, consisting of only a few cells, goes to prove without question that the cells, wherever found, in plants, in the brain of man or insect, or living singly in the water, are the same.

It is not necessary to go into any details in reference to the intelligence of animals; anyone who has had any experience with them knows that they reason in reference to their particular business in life, just as man does. Goats very often meet on very narrow mountain ridges with a precipice or deep gorge on each side. They cannot pass nor can they retreat, as there is no room in which to turn about. Under those circumstances the goats will stand and look at each other thoughtfully for some time and mentally discuss matters; then one of the goats will kneel and lie down very carefully and the other will walk deliberately over him; they never get rattled nor lose their head. You will always find, however, that any species of animal is about as clever as any other in the particular line of business which concerns its existence,

FIG. 53.—Beginning of civilization.—LITERARY DIGEST.

and this we also find to be a fact in the cases of plants, cells, and also of man.

Some claim that particles of organic matter like gravel, stone, clay and other substances possess consciousness, intellect, memory and soul. They claim there can be no distinction between the dead and the living. I am unable to see any foundation for such a claim. We have also on the other hand those who claim that some divine power somewhere not only made all living things but also guides their actions in life. I can see no foundation for such a claim. There may be a higher power in charge of the universe, or there may be intelligence in the atoms or particles of dead matter, but we do not know,—we can only guess. It seems just as likely, if not more so, that there is a separate intelligent and invisible something in the universe, which is able to direct, organize and form matter and force to serve its own purpose, but we do not yet know.

Phrase-making and abuse of language in order to mystify, cover up ignorance or pretend to be clever has been the common practice of late and it has lead to nothing but confusion. Since the microscope discovered for us that a plant or animal was a colony made up of still smaller animals, it seems singular to me that we should pay no attention to that fact, as a cause of the plants' existence, but still insist and claim that either some other being somewhere in the universe is the cause and builder of the plant or else that the dead matter itself is the cause. After we have discovered that man builds skyscrapers and ships, why should we argue that the bricks and stones alone build themselves into a skyscraper and why should others again argue that a divine something somewhere in the universe builds the skyscraper?

We have discovered the builder of plants and animals; we know who he is and what he looks like. Our first investigation, it seems to me, should be whether or not he is an intelligent being. If he is found to be such, then we know how and in what way he is able to build and produce the plants and animals that we see. I believe that we can truthfully say from the evidence at hand that the cell is a conscious and intelligent being like ourselves.

Spotted something that isn't quite right?

Let us know and we'll fix it for future reprints. And as a thank you for your diligence, we'll send you a free Sothis book of your choice.

publisher@sothis.press

This page intentionally left blank

Except of course for that line above. And this one.
And whilst we have your attention, perhaps you'd
like to follow us on Instagram @sothis.press. Go on.

Made in United States
Orlando, FL
04 November 2026

90206689R00277